THE MEMORY ILLUSION

THE MEMORY ILLUSION

REMEMBERING, FORGETTING AND THE SCIENCE OF FALSE MEMORY

DR JULIA SHAW

ANCHOR CANADA

Anchor Canada edition published 2017

Library and Archives Canada Cataloguing in Publication data
is available upon request.

ISBN 978-0-385-68531-3

Cover design: Andrew Roberts
Cover image: Sherbrooke Connectivity Imaging Lab (SCIL)/Getty Images

Printed and bound in the USA

Published in Canada by Anchor Canada,
a division of Random House of Canada Limited,
a Penguin Random House Company

www.penguinrandomhouse.ca

10 9 8 7 6 5 4 3 2 1

To Mark.

'Our memories are constructive. They're reconstructive. Memory works . . . like a Wikipedia page: you can go in there and change it, but so can other people.'

– Professor Elizabeth Loftus

Table of Contents

Introduction

Nobel laureates, on receiving their prize, get the equivalent of a Twitter post explaining what the award is for. Since I learned this, I have spent what is probably far too much time reviewing these 140-characters-or-fewer statements written to describe the profound impact laureates have had on the world.

One of my all-time favourites summarises the work of Seamus Heaney, who won the 1995 Nobel Prize in literature. It explains that he received the award 'for creating works of lyrical beauty and ethical depth, which exalt everyday miracles and the living past'. What an incredible statement. Beauty, ethics and history, drenched with a sense of wonder and captured in just a few words. Every time I read that phrase I smile.

I write these laureate summaries on the little whiteboard I have on my desk as inspiration. I also use them in my lectures, and I try to weave them into my writing. They represent the notion that even humanity's greatest achievements can be explained in plain English. This is an idea that has been echoed by many of history's greats; that for our work to carry significance, we must be able to explain it simply.

I live by this philosophy of explanatory parsimony myself, though of course it does sometimes come at the cost of explanatory adequacy. In other words, when I explain concepts by using analogies, stories or simplifications, I always risk losing some of the nuances of the inherently complex issues under discussion. The subjects I will be examining here, memory and identity, are both incredibly complex, and in a single book I can only hope to scratch the surface of the incredible research being done at the intersection of those areas. But while I cannot promise to capture the whole scientific picture, I do hope to begin a questioning process, one which addresses fundamental queries that have likely been nagging at most of us ever since we first began to utilise the gift of introspection.

Like many others, I first noticed my ability to introspect when I was a child. I remember lying awake for hours as a little girl, unable to fall asleep because I was so engrossed in thought. Lying on the top bunk of my bed I would press the soles of my feet against the white ceiling of my room and reflect on the meaning of life. Who am I? What am I? What is real? While I did not know it at the time, this was when I began to become a psychologist. Those questions are about the core aspects of what it means to be human. As a little girl I had no idea that I was in such good company when I could not figure out the answers.

While I no longer have the bunk bed, I do still have the questions. Instead of philosophising to my ceiling I now conduct research. Instead of discussing who I am with my musical bear, I get to discuss it with fellow scientists, students, and others who are curious like me. So, let us start our adventure through the world of memory at the beginning of all beginnings, where re-search is me-search. Let's ask: What makes you, *you*?

You-ness

When we define ourselves we may think about our gender, ethnicity, age, occupation and the markers of adulthood we have achieved, such as completing our education, buying a house, getting married, having children or reaching retirement. We may also think about personality characteristics – whether we tend to be optimistic or pessimistic, funny or serious, selfish or selfless.

On top of this we likely think about how we compare to others, often directly monitoring how our Facebook friends and LinkedIn connections are doing to see whether we are keeping up. However, while all of these descriptors may be more or less appropriate ways of defining who you are, the true root of your 'you-ness' almost certainly lies in your personal memories.

Our personal memories help us understand our life trajectories. It is only through my personal memories that I can recall the chats I had with one of my most inspiring undergraduate professors, Dr Barry Beyerstein, who taught me critical thinking and often shared lemon poppy-seed loaves with me. Or the talks after lectures with Dr Stephen Hart, who was the first person in my life to encourage me to go to graduate school. Or my mom's serious car accident a few years back, which taught me the importance of expressing my emotions to the people I love. Such milestone interactions carry tremendous significance for us, and help us organise our personal narrative. More generally, memories form the bedrock of our identities. They shape what we think we have experienced and, as such, what we believe we are capable of in the future. Because of all this, if we begin to call our memory into question we are also forced to question the very foundations of who we are.

Take this thought experiment as an example: What if you awoke one morning and could not remember anything that you have ever done, or thought, or learned? Would this person still be you? In thinking about this scenario, we may instinctively react with a feeling of fear. We may have the immediate sense that everything we are could be taken away from us just by taking our memory, leaving us a shell of our former selves. If our memory is gone, what do we have left? We could easily picture it as the premise of a terrifying science fiction movie: 'And then they awoke, none of them knowing who they were.' Alternatively we may look on the prospect with a sense of relief that we would no longer be confined by our past and could start our lives anew, with our basic mental capacities and personality still intact. Or perhaps we find ourselves uncertain, vacillating somewhere between the two viewpoints.

While that kind of dramatic memory loss is thankfully rare in real life, at the same time our memories are subject to an enormous array of errors, distortions and modifications. I hope to shed light on some of them in this book. Armed with science and genuine curiosity, and sprinkling in little bits of my own escapades, I want to repeatedly challenge us to consider the many ways our memory can go awry. But how do we really begin to talk about the complex phenomenon of memory? Let us begin by looking at two of the key terms in memory research.

Semantic memory, also called generic memory, refers to the memory of meanings, concepts and facts. Individuals are often better at remembering certain types of semantic information than others. For example, someone who is great at remembering the dates of historic events might be terrible at remembering people's names; another person might experience the opposite – being great with names but terrible at important dates. Although both of these are types of semantic memory, performance in such tasks can vary considerably between individuals.

Semantic memory works alongside episodic memory, or autobiographical memory. When you remember your first day of university, your first kiss, or the vacation you took to Cancún in 2013, you are accessing your episodic memory. This term refers to our collection of past experiences. It is our personal memory scrapbook; our mind's diary; our internal Facebook timeline. Episodic memory is the mechanism that keeps track of memories that occurred at a particular time and place. Accessing these kinds of memories can be like reliving multisensory experiences. We can feel our toes in the sand, the sun against our skin, the breeze in in our hair. We can picture the venue, the music, the people. These are the memories we cherish. It is this particular memory bank that defines who we are, rather than just the facts we know about the world.

Yet, this episodic memory that we all rely on so much is something many of us woefully misunderstand. If we can get a better picture of how it actually works, we will also gain a better understanding of the circus that is our perceived reality.

Clay with consequences

Once we begin to question our memories, and the memories of others, it seems less surprising that we can often disagree with friends and family about the details of important past events. Even the precious memories of our childhood can actually be shaped and reshaped like a ball of clay. And memory errors are not isolated to those who we may perceive as vulnerable – those suffering from Alzheimer's, brain damage or any other notable impairment. Instead, memory errors can be considered the norm, not the exception. We will explore this potential discord between reality and memory in more depth later.

Similarly, false memories – recollections that feel like memories but which are not based on any real occurrence – are experienced all the time. And the consequences of such false memories can be very real. Believing inherently fictitious representations of reality can affect anything in our lives, potentially causing real joy, real upset, and even real trauma. Understanding our faulty memory processes may therefore help us establish a sense of how we can – and cannot – evaluate the information contained in our memories, and how to use them appropriately to define who we are. This has certainly been my experience.

Over the course of my years of research on memory, I have come to realise that we see the world in deeply imperfect ways. In turn this has given me a great respect for the scientific method and collaborative research – the collective enterprise of science. They offer the best hope of seeing through the veil of our imperfect perceptions to understand the workings of memory. However, even with the wind of decades of memory research in my sails, I must admit that there will likely always be some doubt as to whether any memory is entirely accurate. We can merely collect independent corroborating evidence that suggests that a memory is a more or less accurate mental representation of something that actually happened. Any event, no matter how important, emotional or traumatic it may seem, can be forgotten, misremembered, or even be entirely fictitious.

I now dedicate my life to researching how memory errors can occur, with a particular focus on how it is possible to shape our memories, and the memories of others, moulding actual past experiences to create a fictitious perceived past. What sets me apart from most of the other researchers doing similar work is the nature of the memories I generate. Over the course of just a few friendly interviews I can use my understanding of memory processes to severely distort the memories of my participants. I have convinced people they have committed crimes that never occurred, suffered from a physical injury they never had, or were attacked by a dog when no such attack ever took place. This may sound impossible, but it is simply a carefully planned application of memory science. And while it perhaps sounds a little sinister, I do it in order to help discover how severe memory distortions can come about, an issue which is particularly important for criminal justice settings where we heavily rely on the memories of eyewitnesses, victims and suspects. By creating complex false memories of crime that look and feel real in the lab, I highlight the distinct challenges our faulty memory processes pose for the law.

When I tell people this, they immediately want to know exactly how I do it. I'll be explaining that later in this book but for now let me assure you that it involves no sinister brainwashing, torture or hypnosis. Due to our psychological and physiological configuration all of us can come to confidently and vividly remember entire events that never actually took place.

The Memory Illusion will explain the fundamental principles of our memories, diving into the biological reasons we forget and remember. It will explain how our social environments play a pivotal role in the way we experience and remember the world. It will explain how self-concept shapes, and is shaped by, our memories. It will even explain the role of the media and education in our (mis)understanding of the things we think memory is capable of. And it will look in detail at some of the most fascinating, sometimes almost unbelievable, errors, alterations and misapprehensions our memories can be subject to. While this is by no means an exhaustive study, I hope that it will nonetheless give you a solid enough grounding in the science involved. And perhaps it will leave you wondering just how much you truly know about the world, and even about yourself . . .

1. I REMEMBER BEING BORN

Infant mobiles, tea with Prince Charles, and Bugs Bunny

Why some of our childhood memories are impossible

'I remember being born' – 62 million hits on Google. 'I remember being a baby' – 154 million hits. 'I remember being in the womb' – 9 million hits. People show a huge amount of interest for early childhood and even pre-childhood memories. We all want to grasp for our earliest memories and understand the impact they may have had on us. And perhaps we also want to know just what our memories are capable of during our infancy. Some people, like Ruth who responded to a *Guardian* online question on this topic, are keen to share their earliest recollections:

> I was in a dark, warm place and I felt very secure. I could hear a steady, rhythmic blip blip blip sound (mother's heart-beat) and I was comforted by it. Suddenly something terrible happened and it frightened me (mother's screams, I'm sure). Then the blip blip sound returned and I thought everything was OK. Again the terrible thing happened and this time I knew it would happen again and again. I was terrified! My body was being painfully pulled and squeezed, mother was screaming and I thought something terrible, horrible and awful was happening! Then I came out and the doctor said

> something to me that was friendly and welcoming. I didn't know the words but I got his message! . . . If my mother were still alive I would ask her if there was a large window in front of us with the sun shining brightly through it and if the doctor had a black moustache and was short and fat.[1]

Ruth is one of countless people who claim they can recall their birth. It is also common for people to claim to have memories from when they were babies, apparently remembering what their nursery or crib looked like, or recalling specific events. Over the course of my career I have heard many examples of this. 'I remember all the little airplanes on the mobile above my bed.' 'I remember getting stuck in my crib and being scared because I was caught in the latch!' 'I remember that my favourite toy was a blue musical bear – I would pull the string and it would help me go to sleep. And how could I possibly know that, if not from memory, since we got rid of the bear when I was two?'

When you stop and think about it, it actually is pretty incredible. How could those people possibly remember any of those things at such a young age?

Well, they couldn't.

Your first memory

Everyone has an earliest memory – clearly one of our memories must be the oldest. And, barring a belief in past lives, this memory must be of an event that happened within a knowable time frame – some time between now and when our minds first came into existence. But how can we discern whether the earliest memory we think we have is an accurate representation of something that happened?

When people claim to be able to remember the mobile that hung above their beds when they were a baby, or the hospital room in which they were born, or the warmth they felt inside their mother's womb, they are recalling what psychologists refer to as impossible memories. Research has long established that as adults we cannot accurately retrieve memories from our infancy and early childhood. To put it simply, the brains of babies are not yet physiologically capable of forming and storing long-term memories. And yet many people seem to have such memories anyway, and are often convinced that they are accurate because they can see no other plausible origin for these recollections.

But actually, it does not take much to think of a few alternative explanations. Is there really no other way we could know what our mobile or crib looked like, or that we got caught in the latch of our crib, or that we had a musical bear? Surely there could be external sources for this information: perhaps old photographs or a parent's retelling of events. We might even have memories of objects of personal importance because they were still around much later in our lives.

So we know that at least some of the necessary raw material to build a convincing picture of our earlier childhood can be found elsewhere. When we then place this information into seemingly appropriate contexts, such as a retelling of an early life event, we can unintentionally fill in our memory gaps, and make up details. Our brains piece together information fragments in ways that make sense to us and which can therefore feel like real memories. This is not a conscious decision by the 'rememberer', rather something that happens automatically. Two of the main processes during which this occurs are known as confabulation and source confusion.

As Louis Nahum and his cognitive neuroscience colleagues at the University of Geneva put it, 'Confabulation denotes the emergence of memories of experiences and events which never took place.'[2] This single word describes a complex phenomenon that affects many of our memories, particularly early ones. Of course, in the case of early childhood memories, this definition can fall a bit short: the event may have actually taken place, it is just impossible that our brains were able to store this information at such a young age and present it back to us in a single meaningful memory later on.

Alternatively, the belief that we have early childhood memories of events like birth may be simply due to misidentifying the sources of information. This is known as source confusion – forgetting the source of information and misattributing it to our own memory or experience. Wanting to remember our lovely childhoods, we may mistake our mother's stories for our memories. Or we may meld into our personal narratives recollections told to us by our siblings and friends. Or we may mistake our imagination of what our childhood could have been like for a real memory of what it was like. Of course, memory errors can also be due to confabulation and source confusion working in tandem.

One of the first experiments which demonstrated that we can fiddle with our memories of childhood was conducted by memory scientists Ira Hyman and Joel Pentland at Western Washington University in 1995.[3] Their 65 adult participants were told that they were taking part in an experiment investigating how well people could remember early childhood experiences. They were told they would be questioned about a number of events which they had experienced before the age of six, details of which had already been provided by their parents through a questionnaire. Finally, they were told that accuracy of recall was paramount.

But of course this was no regular childhood memory study. The researchers did not just want to see how well the participants remembered true events – they wanted to see how well they remembered events that had never actually happened. Among the true accounts obtained from the participant's parents they had hidden a false account they had made up themselves: 'When you were five you were at the wedding reception of some friends of the family and you were running around with some other kids, when you bumped into the table holding the punchbowl and spilled the punchbowl on the parents of the bride.' Appropriately, the study is frequently known simply as the 'spilling the punchbowl' experiment.

It is easy to picture this event – it's both emotional and plausible. We all know what weddings look like in our particular cultures and countries. We all know what a punchbowl looks like, or at least what it might look like. We all know that weddings are generally formal events, so we likely picture the parents of the bride as an older couple dressed up for the occasion. It is easy to picture ourselves running around in this situation at the age of five. And, as it turns out, it is even easier to picture all this if we imagine the event happening for a few minutes.

Each participant was asked first about two true events which the researchers had learned about from the participants' parents, and then they were asked about the fake punchbowl incident. After giving participants basic information for each memory, the researchers asked them to try to form a vivid mental image of the event in order to access the memory. They asked them to close their eyes and imagine the event, including trying to picture what the objects, people and locations looked like. The researchers had the participants come back three times, each visit a week apart, and repeat the process.

What they found will astonish you. Just by repeatedly imagining the event happening, and saying out loud what they were picturing, 25 per cent of participants ended up being classified as having clear false memories of the event. A further 12.5 per cent could elaborate on the information that the experimenters provided, but claimed that they could not remember actually spilling the punch, and were therefore classified as partial rememberers. This means that a large number of people who pictured the event happening thought that it actually did happen after just three short imagination exercises, and that they could remember exactly how it happened. This demonstrates that we can misattribute the source of our childhood memories, thinking that something we imagined actually happened, internalising information that someone suggested to us and spinning it into a part of our personal past. It is an extreme form of confabulation that can be induced by someone else by engaging your imagination.

As an aside, besides being an amazing researcher who has contributed greatly to our understanding of false memories, Ira Hyman is a complex character, and instantly likeable. While we are talking about him, here is a quick multiple-choice quiz. Complete the sentence: Ira Hyman . . .

wrote his first academic publication about the Beatles.

has danced in a ballet.

hates pickles.

All of the above.

Of course, the answer is 'all of the above'. And we love him for it.

Super short-term

Let's back up and talk about the neuroscience of memory and exactly why early childhood memories are so prone to distortion in a physiological sense. When scientists talk about memory maturation – how our memories change as we age – they typically talk about changes in short-term memory and long-term memory separately. Short-term memory is a system in the brain that can hold small amounts of information for short periods of time. *Really* short periods of time – only about 30 seconds or so. For example, when we go to remember a phone number and repeat it to ourselves over and over until we dial it, in what is known as the phonological loop, we are using our short-term memories.

This system cannot carry much of a memory load. Since a seminal paper published in 1956 by George Miller from Princeton University,[4] which also happens to be one of the most cited psychology papers of all time, the number of items we are said to be able to hold in working memory at once is seven plus or minus two. In other words, depending on our unique memory abilities and our mental state at the time, our capacity can be diminished to only holding five pieces of information or increased to holding nine. This variability is sometimes noticeable: when we get really tired many of us will find our short-term memory seems to all but disappear.

While Miller's magic number, seven, has been questioned – according to a 2001 paper by Nelson Cowan from the University of Missouri[5] the number of items we are able to hold may really be just four – the same principles remain; we can only keep a few things in short-term memory at any one time, and we can only do so for about 30 seconds.

A concept that often comes up in discussions of short-term memory is that of 'working memory'. This term generally refers to a larger theoretical construct that has to do with how we flexibly keep information in mind while we do things like problem-solving – short-term memory is generally considered a type of working memory. Conceptual differences between these terms and the way they are used are often incredibly important for researchers, but for the sake of this discussion, I'm going to use them interchangeably.

Christian Tamnes and his fellow researchers at the University of Oslo in Norway[6] examined the maturation of working memory in people between the ages of 8 and 22. In a paper they published in 2013 they found that changes in specific parts of the brain were related to improvements in working memory. In particular, they showed that maturation of the so-called fronto-parietal network in the brain was responsible for short-term memory development. The research showed that short-term memory is closely related to our ability to use our higher level thinking (frontal lobe) in harmony with our senses and language (parietal lobe), and that this ability improves with age. The more the relationship between these parts of the brain develops, the better we become at keeping items in short-term memory.

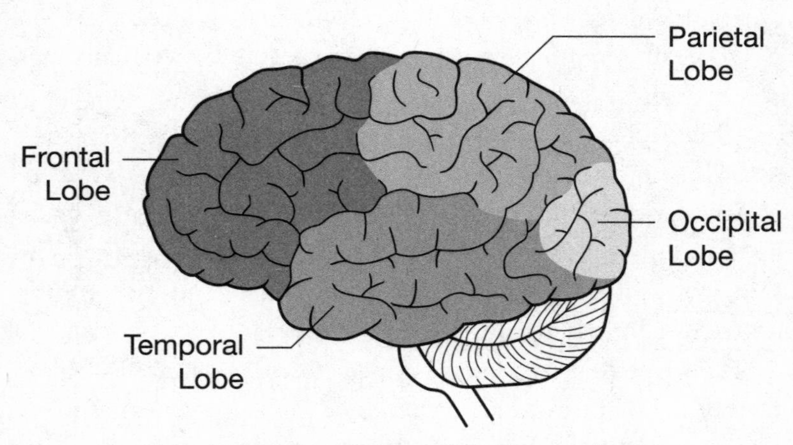

The four major sections of the human brain

That sounds very neuroscience-heavy, so let me break it down. Our brain is divided into four major sections. The parietal lobe, which sits right at the top of the brain, is responsible for integrating sensory information and language, which is necessary for short-term memory. The frontal lobe is the section that sits at the front of the brain, behind the forehead. This part of the brain is responsible for higher cognitive functions such as thinking, planning and reasoning. The prefrontal cortex, the very front part of the frontal lobe, is assigned particular credit for complex thinking, and is associated with abilities such as planning complex behaviour and decision-making.

The prefrontal cortex is a part of the brain that used to be severed in some individuals who presented with severe mental illness, in a procedure known as a prefrontal lobotomy. These crude interventions, which were essentially completed by shoving an ice pick through the patient's eye socket and into the brain, were known to severely affect the patient's personality and intellect. This was considered justified at the time because it was thought to diminish the symptoms of their illnesses. Perhaps this was the case, but only in the sense that the operation typically made those who underwent it into zombies with virtually no personality whatsoever. Prefrontal lobotomies were conducted on many thousands of patients in the US, the UK, the Nordic countries, Japan, the Soviet Union and Germany, among others. The technique was first reported by Egas Moniz in 1936 – he surprisingly received a Nobel Prize for its discovery[7] – but was generally abandoned in 1967, when psychiatrist Walter Freeman killed one of his patients.[8]

Who would have guessed we need such a big network to store such a seemingly small amount of information? Of course, as discussed in Chapter 2, in order to perform even small memory tasks we need to be able to do a tremendous number of things at once – we need to be able to perceive many things simultaneously and sort through them, and we need to be able to integrate that information into our existing memory schemas so that we can understand what it is we are seeing or remembering.

Bringing the discussion back to our early childhood memories, infants and children have been shown to have some short-term memory capability, albeit less than adults, and their memory strategy generally seems to be different – not so much in terms of the basic capacity of short-term memory (although there has been some debate about this over the years) but more in terms of how they approach their environment.

We've already mentioned that short-term memory can hold a certain number of items at any one time. And an item means different things at different times. Let's take a phone number again. While you could try to remember each number individually, seven-five-three-eight-nine-six-zero, it's easier for you to chunk the numbers together; seventy-five, thirty-eight, ninety-six, zero. By doing that you have just diminished the number of items from seven to four, making it considerably easier to keep the number in your short-term memory.

The use of the technical term 'chunking' for grouping things together when performing a task was coined by George Miller,[9] the same man who brought us the paper on the magical number seven. The word is really referring to our ability to apply higher level cognitive processes (hence the importance of the prefrontal cortex) to what we consider a unit in our environment. By using our amazing ability to connect things, our brains can actively or passively organise information into pieces.

For example, if I say 'Starbucks' to you, you know that I mean a multibillion-dollar behemoth of a corporation that started in Seattle. Or, you know, coffee and free wi-fi. What this means is that you already have a representation of 'Starbucks' and what this concept entails in your brain. Thus, in memory processing terms, this counts as one unit of information, rather than the countless different items that you would have to hold in your short-term memory if I just gave you the isolated concepts associated with Starbucks; green, mermaid, coffee, wi-fi, comfy chairs, baristas, venti, grande, tall, latte, muffins, frappuccinos, America, misspelled names on cups . . . you get the idea.

The same goes for the rest of our world. The more we can group ideas or concepts together into chunks, the more impressive our working memory becomes. This is one of the abilities that improves as we get older; as we come to have more experience interacting with and interpreting the world around us, we get significantly better at chunking.

This means that we are better at holding things in working memory in adulthood than in childhood, and we are better in childhood than in infancy, since in our early years we are less able to process stimuli simultaneously, never mind to consolidate them into more permanent memories that can later be accessed in adulthood.

But what about long-term memory? First of all, while short-term memory is indeed very short-term, I should clarify that long-term memory is not necessarily very long-term. What memory researchers mean by 'long-term' is often anything that is kept in the memory for longer than 30 seconds (although, once again, researchers argue about this). However, the term also encompasses memories that we have until we die – including our episodic memories of events and our semantic memories of factual information. And the research on the kinds of long-term episodic memories that last days, years, or even a lifetime has come up with some fascinating results.

Childhood amnesia

Early childhood recollection is one of the most researched areas in the world of memory science. Researchers generally agree that the magic age at which we can begin to form memories that last into adulthood is 3.5 years of age, although some, such as Qi Wang of Cornell University,[10] argue this figure is likely to depend on the individual and can be anywhere between 2 and 5 years of age.

Why? Because in addition to necessary brain structures being underdeveloped, before the age of three everything is new, exciting and unfamiliar. We don't know what is important, and we don't have the structure – and the language – to make sense of the world, never mind the cognitive resources necessary to process it. Because young children and infants don't properly understand or discriminate they don't have any framework for understanding what they should be trying to remember and what they should be forgetting.

This results in a lack of the ability to form early childhood memories that last into adulthood, a phenomenon called childhood amnesia (or infantile amnesia). It is a phenomenon we have known about since 1895 when psychologist Caroline Miles first coined the term.[11] In her research she found that most people's earliest memories were between the ages of two and four. Our understanding of what this means and why it is the case has become significantly more refined since then, but her age estimate was pretty spot on. This is particularly amazing since the notion of false memories, inaccurate pseudo-memories of entire events that never happened, was not to be properly researched or understood for another 70-odd years – when researchers like Elizabeth Loftus came around and revolutionised how we think about memory malleability.

I am not saying that young children do not have memories – they do. Just not memories that generally last into adulthood. From the time we are newborns we can remember simple shapes and colour combinations for up to a day. They are even influenced by the kind of emotion these shapes are paired with; in a 2014 study, Ross Flom and his colleagues in Utah showed five-month-old babies geometric shapes – squares, triangles, circles – at the same time as exposing them to human faces that were either smiling, neutral or angry. This meant that they associated, say, circle with happy, or square with neutral. When tested shortly after exposure, infants were best at remembering the 'happy' shapes. The next day, however, they were best able to remember the shapes that were shown alongside a neutral face. How do we test babies' memories? We measure how long infants look at things. Infants have a preference for new objects, which means that if they remember an object they are going to spend less time looking at it. The results of this study mean that not only can infants remember things for at least a day, which of course counts as long-term memory, but their brains also process and store information about the emotion that was attached to an experience.

From the ability to remember things for a day as an infant, memory capability then increases quite quickly, as two-year-old children can remember some of the events they experience up to a year later. This is why my two-year-old niece may remember who I am if there is a relatively short time between visits, but has trouble remembering me if I don't see her for a year. It explains why we have all experienced this kind of scenario: 'Remember Auntie Julia?' . . . 'No?' . . . 'She gave you that beanie baby when you were little!' *Sympathetic look in my direction.*

We know that parts of the brain responsible for long-term memory, including part of the frontal lobe and the hippocampus, begin to grow at around eight or nine months,[12] so before this it is impossible for infants to have any memories that exceed about 30 seconds. According to Harvard professor Jerome Kagan, one clue that children start to develop memory at about nine months is that this is typically when they become less willing to leave their parents. Being able to miss their mothers is taken as a sign that the infants have a memory of their mother having just been present, and notice when she leaves. According to an interview Kagan gave in 2014 to ABC News: 'If you're five months old, it's out of sight, out of mind. You're less likely to cry because you just forgot that your mother was ever there, so it's not as frightening.'[13]

But whether these memories last into later years is a different question, one that has been addressed by Eunhui Lie and Nora Newcombe at Temple University in Philadelphia. In research they published in 1999,[14] they tested the ability of 11-year-olds to recognise pictures of former classmates from their preschool years. Each child was shown a series of pictures of 3- and 4-year-old children, among which were some images of children they had gone to school with 7 years earlier. Most of the children did not recognise any of their former classmates. And if 11-year-olds have a problem with this task, what hope do adults have 20, 30 or 60 years later? Unless we went to school with the same children into our later years, or remained friends with them into adulthood, it seems likely that we would also be hard-pressed to remember any. And yet, we will have spent *years* with those children. These are not lost memories of short encounters with strangers. No, these are lost memories of years of interactions with the same individuals.

Luckily, long-term memory capabilities develop quickly as we age, both in duration and complexity, as we increasingly understand how the world around us works and what we should consider important. The basic foundations of long-term autobiographical memory are established within the first few years of life, but the main structures involved in memory (the hippocampus and related cognitive structures) actually continue to mature well into early adulthood. This finding has contributed to the notion of an 'extended adolescence' that lasts all the way to the age of 25, since the brain continues substantial maturation until at least this age.

So we can come to appreciate the reality and necessity of

childhood amnesia when we realise that baby brains are just half-baked, unfinished. Not ready for playing in the big memory league.

Baby brains

So big, yet so undeveloped – cute squirmy babies with proportionally giant heads hold a world of potential. Fatty brains that need to become fatter (your brain is actually about 60 per cent fat), which are the most complex structures in the known universe, and which contain the makings of who we will become.

As just mentioned, in our first years of life our brains undergo absolutely massive physical changes. Wanting to know exactly what these changes look like, a team led by Rebecca Knickmeyer at the University of North Carolina used high-tech neuroimaging to take a peek into the brains of 98 children,[15] a number of whom they were able to follow from the ages of two to four weeks right through to two years. In this research, published in 2008, they placed the children into what is known as a structural MRI – a magnetic resonance imaging machine – which can produce a 3D image of the physical structures of the brain. It's really the stuff of science fiction and I would encourage anyone who is eligible to participate in local neuroimaging research – find a local research centre and you may get the chance to look into your own brain. I have done it myself and, of course, I immediately made the resulting image my Facebook profile picture. I have been told I have sexy ventricles.

Back to baby brains. What the researchers found was astonishing. Baby brains increased in total volume by 101 per cent in the first year, and by an extra 15 per cent in the second. That means they more than doubled in size. Broken down by when the MRIs were taken, the baby brain at two to four weeks of age is only about 36 per cent of the final adult volume, 72 per cent at one year of age, and 83 per cent of the final adult volume by two years. If we extend this developmental timeline beyond this initial study, according to another study by Verne Caviness and a team at Harvard Medical School,[16] by the age of 9 the brain reaches about 95 per cent of the adult volume, and it is not until about the age of 13 that our brains reach their full adult size. This increase in brain size coincides quite nicely with the ages at which we start to be able to remember more.

But while tiny baby brains undergo rapid growth, they also face massive neuronal (brain cell) pruning. Pruning means that individual neurons disappear. This process begins almost from birth, and finishes by the time we hit puberty. According to Maja Abitz[17] and her team, adults actually have whopping 41 per cent fewer neurons than newborn babies in important parts of the brain that play a role in memory and thinking, such as the mediodorsal nucleus of the thalamus. If you were to see this pruning process without knowing what was going on, you would almost certainly assume with devastation that the poor human you were observing was about to die a horrible brain death – all those beautiful, galaxy-like, neurons just disappearing forever. But, rest assured, things are meant to happen this way: with great brain growth comes great pruning. It is the process whereby our brain becomes more efficient. Our brains grow, and optimise. Grow, and optimise. Grow, and optimise. So, while the overall volume and size increase, the number of neurons actually decreases, to make way for only the most important and lasting information.

As brains lose neurons but grow in size, they also seem to change the way they make connections between neurons. As described in Chapter 3, neurons are the cells in our brain that process and transmit information through electrical and chemical signals. The connections between them, known as synapses, are often thought to be a reflection of learning processes such as those that allow our working memories to chunk information. Synaptogenesis – the formation of synapses – creates the kind of connections that allow us to form a physical web between associated concepts, as with the Starbucks, green, coffee, barista and wi-fi.

According to research done on this phenomenon by neuroscientist Peter Huttenlocher[18] from the University of Chicago, there is an overproduction of synaptic connections in infancy, followed by persistence of high levels of synaptic density into late childhood or adolescence. This is followed by synaptic pruning, a process that normally ends around mid-adolescence. This means that we start off in life with many neurons and making an incredible number of connections, which we then retain into childhood. However, as we enter late childhood, our brains start to become better at knowing which connections we need to keep and which are superfluous. From there on until mid-adolescence our brains undergo a sort of spring-cleaning. Sure, when you were five years old you could list all of the dinosaurs, but did you really need all that information? Probably not, says your brain, and erases the connections and neurons responsible for much of this knowledge.

Pruning unnecessary synapses is a crucial step in the learning process, as in addition to forming meaningful connections between related concepts in the brain, we need to be able to get rid of inappropriate ones. We prune any potential networks between Starbucks and unrelated concepts such as, say, yellow, flowers and unicorns. This maximises our efficiency when we are trying to remember what Starbucks is, and our ability to apply this knowledge quickly when needed.

As we grow, the intricate web of unnecessary connections between neurons simultaneously proliferates and is pruned down, so that it can more easily be navigated. We grow a tremendous number of neurons, with many possible connections, then get rid of those neurons and synaptic connections that are used the least in what researchers Gal Chechik[19] and colleagues from Tel-Aviv University call optimal 'minimal-value' deletion. Essentially we go from a cluttered brain to an elegant brain that is optimised for our particular environment, according to our individual learning, biology and circumstances.

So, due to structural insufficiencies, as well as organisational and linguistic deficits, memories of early childhood events cannot last into adulthood. But we have yet to really explore why we often think we can remember those years anyway. It is intuitively easy to understand how having insufficient brain capabilities mean that we can forget things that actually happened, but how can we seem to remember things that didn't happen? Why, in the example from the beginning of this chapter, was Ruth so convinced she could remember being born? She had vivid, detailed, multisensory 'recollections'. She described things she heard in the womb, her emotions and the physical pain she experienced during birth, the doctors and the hospital room in which she found herself. How is this possible?

Bugs Bunny and Prince Charles

To explain this, let's turn to a very clever series of studies on infant mobiles. It is the mid-1990s. We are in Canada's capital city, Ottawa. Psychological scientist Nicholas Spanos is sitting with his research team. They decide to embark on the task of demonstrating that it is possible to generate early memories of things that are not just unlikely, but impossible. After some discussion, they submit ethics for a study that will go on to rock the foundations of the scientific understanding of memory, and prove that false memories of early childhood events can be easily generated in most people. Tragically, on 6 June 1994 Spanos was involved in a fatal plane crash and so never finished the work himself.[20] However, it was continued by his collaborators, Cheryl and Melissa Burgess, and in 1999 they published the results.

In the study,[21] participants were given a number of questionnaires to complete which were then taken away by a researcher and supposedly input into a computer. The researcher then returned to give the participant feedback, and told every one of them that they had well-coordinated eye movements and visual exploration skills, which the researchers claimed must have been formed right after birth. Participants were also falsely told that these superb visual skills were probably due to their having been born in hospitals which hung coloured mobiles above the cribs of newborns.

This was, of course, an elaborate lie. It was false feedback that the experimenters had predetermined, to give them an excuse to try to dig into the participant's infancy memory banks. The participants were then told that in order to confirm that they had indeed had this experience with a coloured mobile, they would be hypnotically age-regressed to the day after birth, and subsequently questioned about what they remembered.

Age regression is a process in which an individual is said to mentally return to an earlier stage of life, giving them better access to memories formed at that time. It is a psychoanalytic concept that originates in notions posited by Sigmund Freud. It has also been discredited by numerous empirical studies – it just doesn't work reliably as a memory aid.[22] In other words, the researchers lied about both the premise of their study, that it was about visual skills, and the effectiveness of their supposed memory retrieval tools.

Yet, despite using these discredited memory retrieval techniques in their research, Cheryl and Melissa Burgess found that participants seemed to recall a high number of details about the time they had supposedly been regressed to. In fact, 51 per cent of participants claimed to be able to remember the coloured mobile they had been told about. And, much like Ruth from the beginning of the chapter, many participants who did not remember the mobile recalled other things such as doctors, nurses, bright lights, cribs and masks.

What was shocking was that almost all the participants who had these pseudo-recollections claimed that they were real memories as opposed to fantasies. The researchers had successfully generated false memories of an event that they had made up, from an age when the brain is physically incapable of forming such long-term memories. That meant that participants had been led to confabulate an entire memory out of thin air – they had generated *impossible* childhood memories.

Another researcher who wanted to get participants to believe the impossible was Kathryn Braun at Harvard Business School. In 2002, Braun and her colleagues[23] conducted an elegant and very simple study that subtly manipulated an experience that many North American children share – a trip to a Disney resort. In what was an interesting crossover exercise between memory research and business, the team wanted to know whether an advertisement could induce a partial false memory.

In their first study, the team asked participants who had previously been to a Disney resort as a child to read an advert that suggested they must have shaken hands with Mickey Mouse when they were there. As the researchers had predicted, those who had read the advert were more confident that they had indeed personally shaken hands with Mickey Mouse than those who had not read the advert. In the second study, the team asked different participants to read a different advert for a Disney resort, this time one that suggested they had shaken hands with

Bugs Bunny. Again, the advert increased confidence that the event had actually occurred.

While technically the former experiment could not rule out the possibility of having drawn on real memories, the latter seems more definitely to have convinced participants of an impossible event; Bugs Bunny is a Warner Bros character, so he would have had no place at a Disney resort. It seems that even something as subtle as brief exposure to advertising can manipulate our precious childhood recollections.

This research was important in demonstrating that we can manipulate or confabulate small moments in our lives that are linked to real events, such as a trip to Disneyland. To many, however, this seems a rather mundane snapshot – a false memory for a relatively trivial situation. The question then becomes: Can we do the same with more complex or serious events?

This was the exact question that psychological scientist Deryn Strange asked. She wanted to know whether we can implant false memories of complex events, including absurdly impossible events. Working at a research lab in New Zealand in 2006, Strange and her colleagues[24] conducted an experiment with six- and ten-year-old children as participants. The children were given four images: two photos depicting actual events that they had experienced and two showing doctored photos of events they had not experienced. What Strange wanted to know was whether the plausibility of an event affected the likelihood of the children accepting it as true. She therefore gave the children both a 'plausible' doctored photo of them on a hot-air balloon ride, and an 'impossible' doctored photo of them having tea with Prince Charles.

After interviewing the children about the events three times over the course of three weeks, Strange found that a large number of them, 31 per cent of the six-year-olds and 10 per cent of the ten-year-olds, came to accept the false events as true and generated elaborate details about their experiences. While age seemed to matter – with younger children being more susceptible to generating false memories than older children – plausibility did not. Children came to believe that they had had tea with Prince Charles at the same rate as they came to believe that they had gone on a hot-air balloon ride.

It seems that we can generate convincing memories of false childhood events, even impossible ones, with very little effort indeed.

An idiot whispering like a drunken man

We are beginning to see ways in which memory can be treacherous and wrong. However, anecdotal evidence clearly shows that we are not always fooled – sometimes we realise that our memories cannot be true. When he created the False Memory Archive, artist and Wellcome Trust Engagement Fellow A. R. Hopwood asked the public to anonymously submit a 'false' or 'non-believed' memory . (He went on to work closely with a number of psychologists to develop a series of art works in response to research into false memory for an exhibition that toured the UK from 2013–14.) Here is an example of a false memory reported in the archive:[25] 'I was born in 1979 in Australia, and in 1980 we moved back to the UK to Coventry in the West Midlands, and I grew up there. I have a memory of sitting in a pram beside the construction site of the new Coventry Cathedral, and it's about half built, with scaffolding everywhere. My mum is there, wearing a long green dress.'

There is nothing particularly unusual about this story. It all sounds plausible enough. There are a number of important details mentioned, including a visual memory of what the mother was wearing. It is also terribly mundane, which makes it feel like an unusual event to have made up or imagined. This image of normality, however, is shattered when the rememberer goes on to say: 'The new cathedral was built between 1951 and completed in 1962, 17 years before I was born.'

This particular rememberer reportedly first questioned the memory when they returned to the cathedral, which reminded them of the memory. They then decided to check its accuracy. They suggested that the reason for the false memory was probably that they knew that the main cathedral, along with much of the rest of Coventry, was destroyed during World War II. They would therefore also know that it had to be rebuilt, and that this would most likely have included scaffolding. Together this made it easy to create an image, an impossible false memory, in which all these elements featured together.

In a seminal paper published in 1975,[26] John Flavell and Henry Wellman at Minnesota University coined the term metamemory to refer to an important faculty we possess that likely plays a part in this kind of memory self-correction. Metamemory is an individual's awareness and knowledge of their own memory. It includes our knowledge of our own memory capacity, as well as an understanding of strategies that can improve our memories. And it also includes our ability to monitor what we can accurately remember, and to analyse our memories to confirm their plausibility.

Thus, when we catch ourselves and realise that a memory we have is actually false, we are utilising our metamemory. It is also metamemory that generally allows us to distinguish between things we have imagined, things we have observed and things we have actually participated in – although as we have seen, this ability can be hijacked and misled to generate memory illusions. Without metamemory helping us identify the strength of our memories, assessing our memory plausibility and generally checking in on our memory abilities, we would likely be floating somewhere between reality and imagination at all times. It is why healthy adults don't *always* think that what they imagined was real, and generally have a good grip on what did and did not happen to them.

Here is another example from the False Memory Archive, of a woman who thought she had a real memory until her metamemory kicked in: 'I was in an apartment. Four women were playing cards. The sky outside the window was dark. The curtains were an orangey plaid. The women were smoking; I remember the bluish smoke curling in tendrils toward the light over the table at which they played. One of the women said, "I think the baby's coming!" Subsequently, she was rushed to the hospital.'

Details about the weather, the arguably inconsequential curtains, and the incredibly vivid description of the cigarette smoke make this a compelling account. If we were told this story, it is highly likely we would take it at face value, perhaps even thinking that the teller had a fantastic memory since she was able to recall so many little details. It sounds impressive, until she goes on to state: 'Now, the way I know that this memory is false (even though it is as clear today as when I was a child) is that the baby to whom she gave birth an hour later was me.' The rememberer does not explain where this fabrication came from, but we can think of many possible sources, from her mother's telling of the story, to pure imagination.

If you are questioning the notion that something from *before* birth could even be called a memory, you need only talk to people who believe in the supernatural or past lives. To a memory researcher, what matters is that the person themselves classifies a recollection as a memory. They feel like it happened, and they feel like they remember it, even if it couldn't be true. If we look at our adventure through the world of research on false memories so far, be they false memories of meeting Bugs Bunny at Disneyland, remembering a mobile seen the day after birth, or remembering birth itself, all of these events share the commonality that they are impossible but feel real.

Metamemory is by no means flawless, however, and in trying to make sense of false memories, we can spin further narratives and come up with excuses so as to make things fit, as in this example, once again taken from the False Memory Archive:

> I went to study art history as an undergraduate for four years, carrying the pleasant memory of having [already] experienced a Michelangelo in all its glory, sublimity and sculptural grandeur [from having seen David on a visit to Florence as a child]. I found out that the V&A in London had a replica of the sculpture. When I went to see it, I was surprised to see how anorexic it looked compared to my memory. The one I thought I saw was much grander – [I thought] perhaps it was because I remembered looking at it from a child's point of view when I was fascinated by its strength, immaculate stone quality and dominating presence. I called my dad and told him of my disappointment and to my dismay (that I feel till today) I found out that we never went to Florence and that I never saw Michelangelo's David.

Sometimes the realisation that something cannot have occurred only comes in the light of new evidence that contradicts our previous beliefs. Most of us are not constantly critical of our memories; our metamemory may not be on guard and so lets pieces of fiction slip past it. In these cases, it is only when we actively engage our metamemory again, usually when we see that a memory is unlikely or even impossible, that we can hope to rid ourselves of the false memories that have slipped in undetected. Metamemory is a beautiful thing that can help us decipher fact from fiction, but it too has its flaws.

The formative forgotten years

Before we leave childhood memories behind, I want to make one thing clear. This research absolutely does not suggest that just because we cannot remember them, early childhood events are unimportant. Our earliest years are, of course, fundamental for brain, personality and general cognitive development. According to a 2012 review of the long-term repercussions of adversity experienced in early life, by medical doctor Jack Shonkoff and his colleagues,[27] experiencing adversity, even at an age before we can consciously remember it as adults, can have lasting effects. As they put it, 'Early experiences and environmental influences can leave a lasting signature on the genetic predispositions that affect emerging brain architecture and long-term health.' It is amazing and strange to think that the years which are possibly our most formative are also those which we remember the least.

2. DIRTY MEMORIES

#thedress, time travellers, and the good old days

Why to remember is to perceive

I was recently in San Diego, California, and found myself wandering around the absolutely stunning Balboa Park, lined with huge palm trees and dramatic cliffs. As a reprieve from the sweltering heat, I decided to go inside their science centre. There, I stumbled upon a magic show that promised to combine perceptual science, physics and psychology. I sat down in an audience of excitable children and tired-looking parents. To be honest, I was a bit worried about what I had got myself into. Then the magician, Jason Latimer, walked out on stage and began his show.

I saw him walk through a solid mirror. I saw him form water into spheres with his hands. He put boxes that were exactly the same size into one another. He hung his coat on a beam of light. And to this day I have absolutely no idea how he managed any of this. During the performance, he consistently explained that there was nothing magical about what he was doing, rather he was using his understanding of science to fool our perceptual expectations. Latimer wanted to make us question our understanding of reality, to appreciate the capacity of science to make the seemingly impossible possible. Despite my initial concerns, I could see why they had let a magician into a science centre.

Even when we knew exactly what trick he was about to perform, when he actively gave us a heads-up, I am confident that none of us could figure out how the illusions worked. This is important. Even when we know that they are false and are even expecting them to happen, visual illusions often remain incredibly compelling. And even though I know that my perception influences my experience of reality, and that my memory can be very flawed as a result, it usually still feels incredibly real.

The creation of new memories relies on the raw data of our perceptions, and while our perceptions may always feel accurate, we know that in truth they are not. Even if I know that what I am seeing on stage is an illusion, and am able to explain what I have seen to myself in those terms, that illusion still has all the appearance of reality. And in other situations where our perceptions have been led astray we may not have the advantage of knowing that this is the case, meaning that we accept what we perceive at face value. In this chapter we'll look at some of the ways in which perception and memory interact, and the possible flaws that can therefore be built into a memory right from the moment of its creation.

#thedress

'White and gold.'

'No! Black and blue!'

'No, it's *obviously* white and gold!'

In early 2015 the media was ablaze with this debate. There was a photo of a dress, and some people saw it as being blue with black lace while others saw it as being white with gold lace. Team black and blue and team white and gold dominated many of our Facebook feeds, with friends often becoming quite vocal, convinced that those on the opposing team were colour blind, stupid, or simply lying. Celebrities overloaded the Twittersphere, on teams #whiteandgold or #blackandblue. Dressgate had gone viral.

Aside from being a great way to waste ten minutes squabbling with incredulous colleagues, that photo of #thedress actually says a lot about how our perception works, and how it can mislead us. Such a dramatic difference in perception seems impossible, as though it is some kind of trick. How could different people possibly see almost opposite colours in exactly the same photo at exactly the same time?

Apparently I wasn't the only one who wanted to know more about what was going on here. Three separate papers were published in 2015 on 'dressgate'. One of the scientists fascinated by the phenomenon was Bevil Conway,[1 2] Associate Professor of Neuroscience at Wellesley College, who claimed: 'This is one of the first [documented] instances, if not the first, of people looking at exactly the same physical thing and seeing very different colours.' He went on to say, 'These three papers are the tip of the iceberg. The dress is going to continue to be a very important probe for understanding the fundamental problem of how the brain turns data into perception and cognition: how do you turn the stuff that hits your senses into something actionable, a perception or thought?'

What the papers found was astonishing to some, but might seem quite obvious to others. In the first study, conducted by Karl Gegenfurtner and his colleagues from Giessen University in Germany,[3] the question was simply how most people interpret the colour of the dress. They studied and recorded exactly how a group of people saw the dress, in order to discover which interpretation was most common. They found that rather than having only two possible combinations, participants saw the images on more of a colour continuum, with some seeing the main combinations (black and blue / white and gold) as darker versions of those colours and others seeing more pastel shades. This suggested that there could potentially be other more nuanced teams – team #blackandpastelblue or team #beigeandgold – but it did little to explain why these differences existed in the first place.

Addressing the *why* at the heart of the issue, a team from the University of Nevada, spearheaded by Alissa Winkler,[4] conducted a study to examine whether perhaps the phenomenon known as colour constancy was one explanatory mechanism here. Colour constancy refers to the way our vision compensates for differences in lighting to estimate what colour things 'really' are. It is what allows us to know what colour something is when it is outside exposed to bright sunlight or inside a dimly lit room, and to constantly interpret that colour as the same throughout the day – although the light wavelengths hitting our retinas may change considerably.

Through their research, Winkler and her team actually found an entirely new way of conceptualising colour perception. They discovered a blue–yellow asymmetry, which means that surfaces are far more likely to be perceived as grey or white when the true colour of the surface is actually bluish than when it has a yellow, red or green tint. They argue that this is because there is a tendency to interpret blue tones as originating from light sources, such as the sky. In the case of the dress, its colours could be interpreted either as a product of lighting or as the actual colours of the fabric.

What does all this have to do with memory? It's simple, really. The reason we can have perceptual abilities such as colour constancy is not just because of our amazing physiology, but because we possess fundamental memories informing us about how the world works. We implicitly *know* that blue tones originate from sources such as the sky because we experience it on an almost daily basis. We have a tremendous wealth of memories that inform us what things look like and what they *should* look like given certain contextual cues. And we use these memories of our experiences to help us understand the information coming through all of our senses.

This means that team #whiteandgold interpreted the dress as being dimly lit, meaning the blue tones were interpreted as shade, while team #blackandblue interpreted the dress as being better lit, correctly interpreting its colours as a result. Both teams were using a combination of ocular information and their internal memory-based models of the world.

If you, like me, found that you saw the dress both ways at different times, you can rest assured that research also supported the idea that the image is multistable – it can be perceived differently by the same person when viewed on separate occasions. Summarising all the work done by different scientists on the phenomenon, Bevil Conway concluded: 'The dress is a very powerful tool for understanding how the brain resolves ambiguity . . . there's a lot of [scientific] interest in how internal models shape how you experience the patterns of life. We really thought everybody had the same internal model.'

And for all those still wondering, the dress was *actually* black and blue.

The dress started a conversation about how we all see the world differently, even through perceptual systems that often seem universal. Of course, this is happening simultaneously with other types of perception, not only sight. Indeed, there are more senses than we traditionally talk about; the myth that we have only five has long persisted, but this does little to give our bodies the credit they deserve.

In addition to seeing, hearing, feeling, tasting and smelling, we are also constantly processing information about gravity, outside temperature, humidity, internal temperature, the location of our limbs in relation to the rest of us, tiredness, how our internal organs are doing, and muscle tension, to name but a few. And as with the other senses, if something is interpreted incorrectly during any of these many simultaneous processes – each of which is imperfect – it has the potential to introduce error into our memories at their inception.

One view of the way that we perceive the world is laid out by a model called data-driven or bottom-up processing. It posits that our representation of what the world around us is like is driven almost entirely by the information we collect from our primary senses about our environment, and is minimally shaped by our expectations. This is generally a good model of how our perception works, as most of our experiences *must* accurately reflect the world around us – otherwise we would find it impossible to navigate our environment.

As psychologists James and Eleanor Gibson, writing in 1955,[5] put it, 'the stimulus input contains within it everything that the percept has. . . . Perhaps all knowledge comes through the senses'. The 'stimulus input' is the information going into our brains through our senses, and a 'percept' is simply a mental concept that is developed as a consequence of the process of perception. For example, if you are looking at a flower, it is a stimulus input because it is stimulating your brain through your eyes. If you are looking at and paying attention to the flower, you perceive the flower, making it the percept.

In their seminal paper 'Perceptual learning: differentiation or enrichment?', the Gibsons try to show that our interpretation of our senses does not necessarily rely on past experience. In their own words, 'there is no proof that it incorporates memories'. This means that they claim that we can simply experience things, without bringing memory into it. We see a flower as a flower, whether or not we know anything about flowers. We might not call it 'flower', but it enters all of our brains with petals, a stem, and leaves. The bottom-up model mostly represents how we may intuitively feel perception *should* be – an accurate representation of the real world, built from the information our senses give to us about our environment.

I like it on top

Most of the time, however, we will be encountering things within a particular context. And we all have a host of complex memories and schemas – intuitive ideas regarding how the world works. We almost never just interpret an object in isolation, but instead bring memory into our interpretation of the world. When we look at a flower we don't just see colours and shapes, we also know that we are seeing part of a plant, that it is part of a plant known as a 'flower', and that we probably should not eat it. We also know that it exists in space and that it needs to follow the rules of gravity. This ability to interpret relatively basic information and make sense of it is surprisingly complex and memory-reliant.

Here is another example of just how amazing our ability to extrapolate information from very limited input is. Imagine a drawing of a cube. A few lines on a page can represent such an object – the lines being interpreted to represent three dimensions. Although this is our experience of the drawing, in reality there are an almost infinite number of other figures that could be represented by the same set of lines when collapsed but we simply do not think about these infinite alternatives. The reason we can take this simple input and see 'cube' is because we have more than just the lines on a page to go by. Our visual systems have evolved to interpret the stimuli around us. They have been honed by our experience with the natural world which has shaped our understanding and expectations. Therefore, when we see a collection of lines that we would ordinarily associate with a three-dimensional cube we use our top-down system to interpret them correctly, even though from a bottom-up perspective they are simply a collection of lines on paper.

Essentially, we often make educated guesses about the world as we perceive it, based on our past experiences. When you think about it, this is an essential skill for survival. Since prehistory, humans have often had to make decisions based on limited information. If we were to take the time to thoroughly check what *those glimmers that look like lion eyes* were, we probably wouldn't survive to ponder such questions another day. And we certainly do not have enough time to check whether all objects in our environment that look three-dimensional actually are. So we take what we might think of as reasonable interpretive shortcuts. Perception is experienced as a coherent and fluid process only because our brain is constantly making educated guesses, filling in the gaps in information.

Let's take first impressions as an example. What happens when we first meet someone? Do we look at them objectively, taking apart each physical feature to identify whether they are friend or foe? No, of course we don't. In 2013 I published some research relevant to this.[6] I was working with three memory scientists at the University of British Columbia, Natasha Korva, Leanne ten Brinke and Stephen Porter, and we were interested in how bias in perception might affect legal proceedings. In particular we wanted to know what makes someone appear trustworthy to others. Trustworthiness, it turns out, is an elusive concept. When we meet someone we may feel we are making an informed decision about whether or not we can trust them. We may think we are making a bottom-up decision, looking at all available evidence and making a rational decision as a result. This is not so. Not at all.

In our study we gave participants a picture of a 'suspect' along with vignette about a crime they had supposedly committed. Then we allowed the participants to be jurors, deciding whether the person was guilty or not. They based their decision only on the photo, the vignette and a set of evidence. We randomly matched the vignette and evidence with a photo; if participants were being impartial they should have based their decision only on the evidence, not on the photo. The photos had already been assessed in a pre-screen, which asked participants to rate on a scale how trustworthy they found each face.

As we had expected, participants were far harsher on those faces that looked untrustworthy than on those who looked trustworthy. They needed fewer pieces of evidence to come to a guilty verdict for the untrustworthy faces and they were less likely to change their minds when presented with exonerating evidence. Depending on the face of the alleged perpetrator, we got radically different verdicts based on exactly the same evidence. Clearly our participants were relying on something other than the actual evidence, apparently making decisions about a person's guilt or innocence based on their existing biases.

Most of the time the automatic involvement of memory to make assumptions and educated guesses is beneficial; it makes us considerably faster at interpreting the various stimuli around us and such guesses are usually accurate. Let's return to our first impressions of people as an example. Research on what are referred to as 'thin slices', meaning brief observations of others that last from a couple of seconds up to five minutes, shows that people are often quite good at inferring certain kinds of character traits.

Nalini Ambady and colleagues at Stanford University have been conducting studies in this area since 1992[7] and have demonstrated that people are pretty good at correctly guessing someone's sexual orientation, teaching performance, and even their ability to deceive others, based only on such thin slices. Unfortunately for our study on how eyewitnesses were influenced by trustworthiness, this ability translated into biased legal decision-making, a context where relying on intuitions and stereotypes can lead to the wrong people going to prison.

Our memories of previous experiences influence our understanding not only of how we expect people to behave, but more generally of how the world works – gravity, dimensions, possibilities. In the same way that we can be fooled by first impressions of faces, we can be fooled by general perceptual illusions, such as those used by the magician I encountered at the science centre in San Diego. If our perceptions have been in some way misled and we do not realise it, the guessing processes our brain employs to help us understand the world can actually have precisely the opposite effect, and can taint our memories with inaccuracy.

Aroused

Are you aroused? On a scale from 1 to 10, how aroused are you? What do you think increases your arousal?

This may sound like the beginning of a poorly cast, low-budget adult film but it also sounds like some pretty typical memory research. So typical that if you type 'memory' and 'arousal' into Google Scholar, a version of Google that searches only academic sources, you get over a quarter of a million hits.

Before you get too excited, though, I need to get your mind out of the gutter. When a researcher says a participant is aroused, they mean that their heart rate, sweating, pupil dilation or other physiological indicators are relatively increased. And it turns out that our level of arousal plays a major part in our ability to encode, store and retrieve memories.

In one of many similar experiments from the 1990s, Larry Cahill and James McGaugh[8] from the University of California wanted to examine this question of how arousal influences memory. In one of their studies, published in 1995,[9] they assigned participants to one of two experimental groups – they were either given a neutral story or an emotional story. Participants in each group viewed the same series of images, but with different narrations played on tape to accompany them. Both stories showed a mother taking her young son to see his father at work – but in the neutral version the narration explained that the father was a mechanic repairing a car, while in the emotional version the narration described him as a surgeon working on car crash victims.

Two weeks later participants were tested on their memory of the story. The researchers found that participants who had experienced the emotional condition could recall an average of 18 details of the event, while those in the neutral condition could recall only 13. In later experiments with a slightly modified method, it was again found that participants in the emotional condition performed better.

These results seem to make it clear that an increase in arousal is associated with an increase in memory performance. This makes sense if we think back to our most vivid memories, which tend to be of emotional occurrences. It is therefore tempting to jump to the conclusion that more arousal is always better for memory retention. But let's think about that a bit more carefully. If I were to ask my undergraduate students on a test day whether they agreed with this, I am not sure that they would – being over-aroused or panicked can make us go blank and forget information that otherwise would come to us quite easily. The 'Ah! But I *knew* that!' post-hoc insight after we have forgotten something during an exam is all too familiar to many of us. Similarly, being in an under-aroused state, such as feeling drowsy and lethargic, is also not going to get you far in an exam.

So we need a more nuanced understanding of memory in relation to arousal. The Yerkes–Dodson theory of performance can help us with this. Developed in 1908 by Robert Yerkes and John Dodson, the theory suggests that performance on any task will improve as arousal increases up to a certain optimum point. However, beyond that point further arousal will actually worsen performance. The suggestion is that at the extremes, at no arousal or incredibly high arousal, a person cannot perform a given task at all. On a graph, this can be represented as an upside-down U – performance initially tracking up in line with increased arousal, then levelling off and decreasing as further arousal becomes detrimental – which is why the theory is called the inverted-U hypothesis.

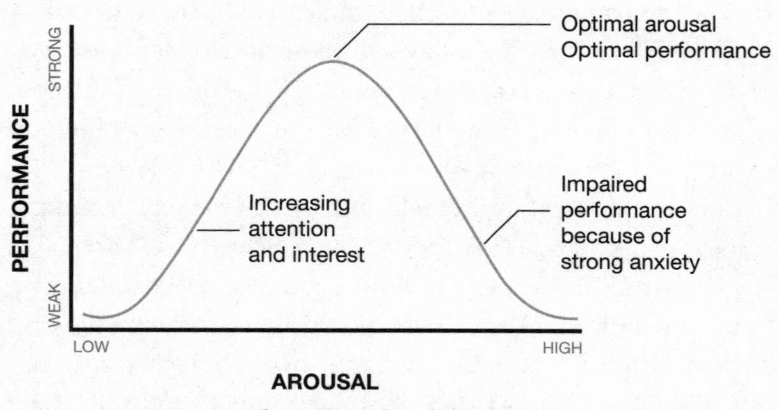

The inverted-U hypothesis

In a demonstration of the inverted-U hypothesis in relation to memory in particular, in 2013 Thomas Schilling[10] and his colleagues at the University of Trier in Germany published a study on how the stress hormone cortisol impacts memory performance. Cortisol is released into the bloodstream when acute stress or arousal activates the HPA axis in the brain – the hypothalamic-pituitary-adrenal cortex. The hormone then crosses into the brain and contributes to the regulation of our stress response, helping to determine how strong and how long it will be.

Schilling's team first asked participants to come in to see 18 pictures of male faces accompanied by a brief description of each person, such as 'he likes to get drunk at parties and then becomes aggressive'. Once the researchers were sure that the participants had learned the associations between the faces and descriptions, they sent them home. A week later, they were asked to come back into the lab. This time they were injected with one of five different levels of cortisol, from none to 24mg. They then had their memories of the associations between the faces and their descriptions tested. The results supported the inverted-U hypothesis, with memory performance increasing up to a moderate level of cortisol and then steadily decreasing after an optimal level was passed.

So the inverted-U hypothesis does seem a good general model for understanding memory performance in relation to arousal. However, there may not be a one-size-fits-all solution. In a summary of the science behind this, published by the American Psychological Association, memory scientists Mara Mather and Matthew Sutherland from the University of Southern California[11] claim that the inverted-U hypothesis does not tell the whole story. They argue this because 'the findings we reviewed here indicate that emotional arousal makes things that are perceptually salient stand out even more and makes any high priority information even more memorable. At the same time, arousal reduces processing of low priority information. This increase in selectivity under arousal is likely to be adaptive in many situations, and can explain why sometimes arousing stimuli impair memory for nearby stimuli and sometimes they enhance it.'

In other words, as our arousal increases, our memory focus generally narrows. We become better at remembering critical information about the incident that made us aroused, but we often become worse at remembering contextual information. If we were present at a bank robbery, for example, we will likely be great at remembering that there was a gun pointed at us, but terrible at remembering much else. Unfortunately, arousal states such as fear do not necessarily focus us on things we will need to remember later. In our bank robbery example, it would be far more prudent to remember the faces of the bank robbers rather than the gun, but according to the much-researched 'weapon-focus effect' our arousal is going to make it difficult to remember anything other than the gun in much detail.

Things are further complicated by research demonstrating that individual characteristics such as age, gender and personality may also play a role in how arousal affects memory – the take-home message is that one size does not fit all when we talk about exactly how memory and arousal are intertwined.

I've saved my favourite association between arousal and memory for last. One way to use this association between arousal and memory to our advantage is to appreciate what is called state-dependent memory, a phenomenon that has been repeatedly demonstrated and validated. What it means is that we generally remember things better if we are in the same state during the recall of a memory as during the encoding of it.

In 1990, Shirley Pearce and her colleagues at University College London[12] demonstrated this in two extremely interesting experiments. The first involved giving both chronic-pain patients and non-patients who were not suffering any pain a list of words to remember. They found that those who reported suffering chronic pain were much better at remembering words associated with pain than any other kinds of words. This is in line with the idea that 'mood congruency' matters; that we are better at encoding and retrieving information that fits with our mood. But this is not enough, as the exact state we are in can change, so Pearce and her team wanted to know whether a temporary state can also influence our memory.

To do this they inflicted pain on some of their participants by asking them to submerge their hands in ice water, while others got pleasantly warm water. If you have never held your hand in ice water for any length of time, this is a surprisingly terrible experience. Right after this water bath experience, the participants were given a list of words to remember. They were then either given another painful ice-water bath, or were asked to put their hands in warm water, before being tested on their memory of the list.

The researchers found that when participants experienced the same state at the time of learning as at recall, they performed significantly better. So those who experienced pain right before they learned the word list performed better if they experienced pain again right before they needed to recall the information. Similarly, those who experienced nice warm water before learning did better at the memory test if they had just placed their hands in warm water again. If we follow this through, it means that if we know we learned or experienced something during a particular type of arousal, by recreating that state we should be able to remember it better. Torturous ice-water baths not your thing? Here's a more pleasant example: if you always drink a cup of coffee right before you study, your memory should be better if you drink a cup of coffee right before your exams.

All this research clearly shows that our stress and arousal levels matter for what we are able to store as memories, and how we are able to retrieve them later on. So our memories can be affected not only by uncontrollable parts of our external environment, but also by largely uncontrollable elements of our internal environment.

Time travellers

Also dependent on our arousal and emotional state is our perception of time. As we all know, the more aroused or excited we are the faster time seems to pass. Popular sayings like 'time flies when you are having fun' or 'this is like watching paint dry' suggest that how much we are engaged by an activity has a clear effect on how we remember it temporally.

Think about it. How long do you think it took you to read the last paragraph? Did it take a long time or a short time? How about assigning a specific value? Ten seconds? One minute? How accurate do you think you are? What do you think your answer is based on – how do you *know* how long it took?

Of course, we do not typically ask ourselves these kinds of questions, and mostly take our perception of time for granted. We too often seem to think that we have some sort of magical internal clock that keeps time for us in a relatively objective manner. However, if we think about those situations in which time seemed to drag on forever during an unpleasant task, or sped by because we were so excited about something, we know that nothing could be further from the truth.

Sometimes referred to as the fourth dimension – an extension of our 3D physical reality – time is something that could be considered a primarily internal phenomenon. It is characterised by linearity, sequentiality and change; by growth or destruction. Our subjective perception of time is known as chronesthesia,[13] and it is studied by researchers from fields as diverse as neurophysiology, psychology and philosophy. And what all of these scientific disciplines have demonstrated is that, perhaps unsurprisingly, memory is vital for our ability to perceive time.

One line of research argues that the way we perceive the passage of time is through our sense of chronology. In other words, we remember the order in which events happened, which then allows us to infer when and for how long an event took place. Obviously to do this, we need to remember what things happened, and in which order. Time is memory; memory is time.

Nobel laureates and behavioural economists Daniel Kahneman and Amos Tversky have done a great deal of work on how we estimate and value time, particularly focusing on things from a memory perspective. They would say that many individuals, particularly those with time prediction problems, engage in the 'planning fallacy',[14] which means they overly focus on what they refer to as 'singular' information, information that is associated with a single task.

For example, if you were a doctor trying to predict how long an Alzheimer's patient will live, the relevant singular information would include how old the patient is, how sick they are, and what their personal medical history is. These are all important pieces of information, but they only really become useful when we place them within the context of 'distributional' information. Distributional information refers to a wider set of information, including *in general* how long, on average, 70-year-old patients with Alzheimer's live. The singular information helps you see how this patient may be different from others, their unique risk factors, while the distributional information helps you use that information to make predictions based on the averages seen in others with those kinds of characteristics.

Distributional information is, of course, reliant on your ability to remember and access the past occasions when you had patients like this, and when you learned that the average life expectancy for Alzheimer's is eight to ten years. Having the ability to contextualise singular information within our distributional information in this way greatly improves our ability to make accurate predictions about how long things will take in the future (or, in our example, how long a patient will survive).

We all have that friend (or family member, or co-worker) who, when they organise events or plan their day, engages in deeply flawed time estimation. The 'Oh, it'll only take me five minutes to get there!' kind of person. We might say such people are 'optimistic' in their time estimates. But we might also say that they are potentially bad at remembering how long it actually took them to do things in the past. They are worse at using their distributional information to ask 'How long does it normally take me to get there?' Sure, Google Maps says five minutes, but that is not accounting for fixing their hair, finding their keys, putting on a coat, going down four flights of stairs, and finding the right doorbell upon arrival. A memory science view of this prediction inadequacy suggests that some people may be late because their memory and time perception systems give them both a poor sense of their past experiences and poor 'prospective' memory – the ability to plan the future based on past experiences.

So, in general how good are we at predicting how long things will take? In a review of research on prospective memory published in 2010,[15] Roger Buehler from Wilfrid Laurier University and his colleagues in Canada looked at research asking individuals to estimate how long particular activities would take them. They found that people were generally optimistic in their estimates, tending to discount past failures to complete things on time, and generally underestimating how long tasks actually took to complete. In other words, we seem to believe that our future selves are going to be superheroes at doing things quickly – new you excels at doing things quickly, even if old you was slow. New you is efficient, old you was lazy.

We can probably all remember times when we've done this ourselves, thinking we will get up early to go for a run, followed by an early breakfast, then finishing off that work by noon, having an effective lunch meeting, catching up on all our emails, visiting the dentist, going to yoga, cooking a five-course dinner, cleaning all the things, going out for drinks with friends, and then heading home for some incredible sex before having a restful night's sleep, all in the same day. When has that kind of day *ever* happened? Yet, how many times do we sit there at night thinking 'Yes, tomorrow is going to be that kind of day.'

Another reason we may be impossibly optimistic in this kind of time estimation is because we may remember how long each of these tasks takes when considered in isolation, but we forget how long it takes to task-switch and move from one task to another. Also, we may forget that generally after a period of exertion our cognitive resources are depleted and need to recharge before we can effectively begin another task. In other words we remember certain elements of how long things take, and discount others.

Further, according to Roger's own research from 1994,[16] and the review of the literature by him and his colleagues in 2010, we only think of *ourselves* as future-time superheroes. When estimating the efficiency of others we are actually comparatively pessimistic, overestimating how long it will take them to complete tasks, and often predicting that they will run into problems that will delay their completion of the task. Researchers in this area have found that this effect applies across different kinds of task estimation, suggesting that our prospective memory abilities as they relate to other people are equally bad in work or personal contexts – we overestimate how much time our friends are going to need to get work done, and how long it will take them to meet us for coffee.

In terms of our autobiographical memories, this means that every time you encode a memory of an event – which will invariably have time-related elements to do with duration and chronology as part of it – you are encoding it through the filter of how you feel, how much is happening that day, and various other biasing factors. Time is not objective, so it is open to the same subjective biases as everything else. And these initial biases, like many of the other perceptual biases covered in this chapter, colour our memories.

Through the telescope

Having an appreciation of time perception is also critical for understanding what is referred to as retrospective timing. This has to do with our sense of the duration of events *after* we have experienced them: estimating how long you were playing a video game, for example.

'Intuitively (without thinking or counting), I have the impression that this game session lasted ___ minutes and ___ seconds.' This was one of the questions that Simon Tobin and his research team from Laval University in Canada asked over 100 participants who were regular gamers, in research they published in 2010.[17] They asked these gamers to come in and play a gaming session in the lab; they wanted to see how good these people were at estimating how long a session of gaming took. They found that when participants were asked shortly after playing, a 12-minute session was generally perceived as proportionally longer than a 35- or 58-minute session. That is to say, the participants were likely to overestimate how long the short session took by a ratio of 1.4:1, estimating that the 12-minute session actually took closer to 17 minutes. They were more accurate with the longer sessions – for

35- and 58-minute sessions many players got the timing almost exactly right. So our retrospective timing of shorter events is considerably less reliable.

Beyond our ability to remember how much time we wasted playing a computer game, our retrospective memory also helps us to remember our entire life timeline. It is useful for timing the duration of individual events, and the duration between events. It allows us to date an event autobiographically, so that we know the relative recency of an event in relation to today, be it earlier today or ten years ago. Retrospective memory allows mental time travel, enabling us to peek into our personal past.

One type of cue we use to help us date memories is known as 'landmarks'. Landmarks are important events such as the assassination of JFK in 1963, 9/11 in 2001, or when Russia invaded the Ukraine in 2014, which people can use to orient themselves along their time continuum – it may be easier for us to chronologically place a personal event if we remember it in relation to these landmark events. For example, I went on vacation to Cuba *after* 9/11, but *before* the Ukraine invasion. This narrows down the time frame somewhat when we are talking about a lifetime of events. These landmarks can also be personal, using events such as a high-school graduation or a wedding as markers. So, instead of pinpointing in relation to historical events, you may remember you went to Cuba after graduation, but before you got married, for example.

Of course, our estimation of landmark events themselves is also fascinating, since we can make mistakes dating them as well. Because the timing of commonly held landmark events is far easier for researchers to verify than personal landmarks, they are a useful phenomenon for studying time estimation across different people. At least that is what a team led by George Gaskell from the London School of Economics thought.

In 2000[18] they published one of the largest investigations of its kind, involving over 2,000 participants who had taken part in a census survey in 1992 which had looked at how good the British general public was at estimating when historical landmark events occurred. They examined two events: Margaret Thatcher's resignation as prime minister of Britain, which had occurred 19 months before the study, and the Hillsborough football disaster, which had occurred 37 months before the study. They wanted to see whether the public could date the events accurately, which they defined as dating the events within one month of their actual occurrence.

The results were fascinating. Overall, only 15 per cent accurately dated the resignation of Margaret Thatcher, and even fewer, around 10 per cent, accurately dated the football disaster. Performance was equally poor across all ages, so it did not seem to matter how old the participants were at the time the landmark events occurred. Instead of accurately dating the events, the overwhelming majority of participants engaged in what is referred to as temporal displacement, or 'telescoping', which means moving things around in time. We have a tendency to do this. In particular, we often remember things that happened more recently as having happened longer ago than they actually did. Conversely, we often remember things from long ago 'as if it were yesterday'.

The telescoping study by Gaskell and his colleagues found that most people reported the event as happening more recently than it actually did (forward telescoping), so closer in time to today. A significant number, however, reported it as being earlier than it actually was (backward telescoping), so being longer ago. For the more recent event, Margaret Thatcher's resignation, which had only happened 1.5 years before the interview, 40 per cent of participants forward-telescoped, while 31 per cent backward-telescoped. The opposite was found for the event that happened over 3 years before – with 29 per cent forward-telescoping and 42 per cent backward-telescoping. Although these effects have been shown in different configurations by other researchers, what they all agree on is that in the time estimation of our memory we have a tendency to push some events even further into the past than they really were, while others we draw closer to today.

Indeed, the bias seems to enter when a memory hits about three years. Things that happened less than three years ago we generally assume were less recent than they actually were, while things that happened more than three years ago feel more recent. While telescoping is due to a complex interplay of memory biases, one reason we might particularly think that things happened more recently than they actually did (forward telescoping) is because landmark memories are often very accessible. We can recall these important life events easily and with a lot of detail, just like memories of things that happened much more recently. We thus interpret this easy access and high vividness of the memory as indicating that it must have happened fairly recently.

Landmarks are an important mainstay of our memory timelines, yet they are riddled with predictable errors. They help us navigate our personal chronology, but it's easy to see how they might prove somewhat unreliable anchors. So, not only are we bad at estimating how long things are going to take, and how long tasks *just* took, but our recollections can play tricks on us that move even important events up and down along the timelines of our lives.

The good old days

Let us think about that timeline of our lives a bit more. Time-travelling through our memories, we may find that some events stick out more than others. If we think about the characteristics that these memories have in common, we may notice that the most vivid are the most emotional, most important, most beautiful or most unexpected events of our lives. We may also notice that our memories cluster. And they often seem to cluster around particular periods in our lives.

This is a phenomenon called the reminiscence bump, and may help explain 'the good old days' and the 'when I was your age' comments. The reminiscence bump means that we do not remember all ages in our life equally. In 2005 a study involving 2,000 participants between the ages of 11 and 70 from the US and the Netherlands was conducted by a team lead by Steve Janssen at the University of Amsterdam.[19] They wanted to answer the question 'What remains in a lifetime of memories?'

Apparently what remains most are memories from between the ages of 10 and 30. The findings of the study supported what others had shown before them – that before the age of five most people report almost no memories. Then, between five and ten, the number of memories begins to increase, hitting a peak for both genders in the late teens. This period of increased reported memories stays quite high until the early twenties, when it begins to drop and then stabilise for the remaining decades. So we seem to retain the most memories of our teens and twenties.

This effect appears to be completely cross-cultural. A study conducted in 2005 by Martin Conway from Durham University in the UK[20] found that it does not seem to matter whether you are from Japan, China, Bangladesh, England or the US – everyone has the same reminiscence bump.

But while the density of memories we recall from this critical time may be the same for all of us, there may be cultural differences in the nature of the memories we have for this time frame. Chinese participants showed a preference for memories that involved groups and events focused on social situations, while US participants generally remembered more self-centred events. Chinese participants reported more memories that focused on themes such as childbirth, interacting with neighbours and colleagues, and intimate relationships, while Americans reported more personal themes such as individual success, frustration, fear, or nightmares. So, our reminiscence bump is the same around the world but with slight variations in the content of our most memorable life experiences according to the culture we live in.

One explanation for the reminiscence bump may be that it is related to our emergence of a real sense of self, which seems to be a largely universal phenomenon. At what age did you form a stable identity? The chances are that if you are a woman, your you-ness first really shone through when you were between 13 and 14, and if you are a man, you probably settled into yourself a bit later, between the ages of 15 and 18. This age range also happens to be the peak of the bump, at least according Steve Janssen's research team.

These are the memories that *defined* us. They are the memories that made us who we are. And, whether or not they have been tainted by perceptual and memory biases, they are the memories we seem to cherish and remember most vividly.

Yet, despite defining us and being so important for our identities, it seems clear that our memories can have inbuilt flaws as a result of the ways our perceptions can be fooled – by visual illusions, our level of arousal, and even from having a poor grasp on the seemingly intuitive ability to sense time. And this is just the tip of the iceberg. Every single one of our perceptual abilities is imperfect. Our vision, our hearing, our taste, our sense of heat, our tactile senses, our vestibular sense of balance, our proprioception of our body in space – every one of these can be fooled.

As the philosopher George Berkeley said, 'esse est percipi' – 'to be is to be perceived'. Only our perception of reality matters. It means that our misperceptions of reality can be placed into our memory systems to be later recalled in spite of never accurately representing more objective reality. While they may generally be close enough to be functionally useful, the truth is that probably every single one of our memories – even the clearest – contains perceptual flaws and inaccuracies, right from the outset.

3. DANCING WITH BEES

Roofies, sea slugs, and laser beams

Why brain physiology can lead our memories astray

So you want to read a book on memory but don't want to hear too much about brain biology? You are not alone. Head on past this chapter then, please. I hereby give you a pass if you don't want to find yourself knee-deep in animal studies, biochemistry, and the history of memory theory. You will still be able to understand the following chapters without any understanding of these topics. However, if you *do* like all of these science-y topics, which really demonstrate to you what memory is, keep reading. And if you have chosen to keep reading, let me introduce you to Kathryn Hunt.

Kathryn Hunt is covered from head to toe in a beekeeper's suit. She is slowly making her way to a hive that is absolutely brimming with *Bombus terrestris* – bumblebees. There are hundreds of them, and she is starting to wonder why she signed up for this. They look cute, being one of the fluffier and rounder varieties of bees, and they are generally not aggressive, but walking towards hundreds of them one cannot help but experience a tinge of anxiety. This is a heart-stopping exercise. Is it even going to work?

Hunt need not have worried – the results of her study were groundbreaking. When she conducted it, she was an eager young researcher at Queen Mary University in London, looking at behavioural decision-making in animals, focusing on whether memory changes in a predictable way as it relates to certain behaviours, rather than just degrading steadily over time. She came from a background of researching how animals are affected by, and interact with, their environments. In this field, known as cognitive ecology, she had previously looked into the food preferences of guppies, and at foraging behaviour in leafcutter ants. It was not until she began to collaborate with Lars Chittka, a leading contemporary expert on bees, that her attention turned towards bumblebees.

Hunt and Chittka wanted to see if they could create false memories in bees. Their research, published in 2015[1] in the journal *Current Biology*, was intended as a demonstration of some of the fundamental physiological processes that are thought to underlie false memory. Bees have a highly evolved social life, a complex communication system, and exceptional ability to learn new information. Their memory seems to work in a broadly similar way to our own, making a study of it also relevant to the science of human memory.

The particular type of bumblebee they used, the aforementioned *Bombus terrestris*, was already known to have a highly developed memory for colours and patterns, including an excellent memory for specific flowers. It had also been shown to have a particularly good capacity for remembering multiple things at once. In short, it has exceptional bee-memory. The researchers wanted to see whether this exceptional memory could be a tricked into becoming a disadvantage for the bees rather than an advantage.

The researchers presented the bees in the study with two unique flowers, one at a time, both of which contained delicious nectar – a flower with black and white rings, and a colourful yellow flower. The bees would therefore have learned to associate these flowers with the nectar they were looking for. During a subsequent test, the bees were given a choice of three flowers: the one with black and white rings, the yellow one, and a new flower that was yellow with black rings. When the bees were given this test minutes after they had encountered the original two flowers they rarely went for the new combined flower, instead correctly picking the flower that had most recently provided them with nectar. They displayed accurate short-term memory.

However, when tested a day or three days later, some of the bees developed a preference for the new 'merged' flower – even though they had never actually seen it during their training, only during the test, and it had never given them nectar. By the end of the study about half of the bees preferred the merged flower over the correct one.

At this point some people might have concluded that the errors in performance meant that the bees had failed to learn which flowers reliably contained nectar, or perhaps had learned and then forgotten. But if the bees had simply forgotten which flowers gave nectar and which did not, they should have selected each of the three flowers presented during the test equally often, rather than showing a marked preference. Instead, Hunt and Chittka interpreted the bees' behaviour as indicative of a false memory having developed – the bees had mixed up their memory of one flower with their memory of the other, leading them to ultimately prefer the mixed flower because it featured both the features they had come to associate with nectar, rather than just one. They had created a merged false memory, and this muddled memory had real behavioural consequences.

All of earth's creatures face similar survival challenges – having to procure food, social networks and mating partners – and research shows that this leads to many overlaps in the cognitive features of insects, other animals and humans. This makes it highly likely that this kind of error is far from being unique to bees. As

Hunt and Chittka put it, '[S]ystematic memory errors may be widespread in animals . . . memory traces for various stimuli may "merge" such that features acquired in distinct bouts of training are combined in an animal's mind, so that stimuli that have never been viewed before, but are a combination of the features presented in training, may be chosen during recall.' Muddled memories are arguably the norm for all kinds of creatures.

For bees, or indeed any other animal, to have evolved a memory which contains such a capacity for error seems hard to believe. After all, being prone to make mistakes which are potentially detrimental to one's survival would hardly be favoured by natural selection. The implication must be that the same systems which leave memory open to such mistakes also confer benefits which outweigh those potential disadvantages. To understand what those may be, we need to get a sense of the bigger picture, and to look in greater depth at the physiology of memory.

Plastic brains

Exactly how we are able to store an experience or thought in the brain has been a question that has fascinated us ever since we first began to entertain the idea that there might be no such thing as a spirit or soul (or that if there is, it might not be an extension of the brain). If that is the case, *all* information *must* be stored physically in the brain. This movement away from dualism (the belief that the mind and body are separate) to monism (the belief that all thinking originates in the brain) has led to a pervasive desire to understand the physical workings of the brain.

While the philosopher Descartes, a famous dualist, believed that the soul and body interacted through the pineal gland, a pea-sized structure located close to the centre of the brain, most scientists today agree that consciousness is not related to an incorporeal spirit, but rather is the result of a complex array of physical systems. Thanks to modern technology, including brain-imaging procedures such as fMRI (functional magnetic resonance imaging) and EEG (electroencephalography), we no longer need to study these systems by dissecting cadavers or examining case studies; for the first time in humanity's existence we can now examine active brains while they are perceiving the world.

Our brains are incredibly malleable and adaptive. They are created for a world of uncertainty and quick decision-making, having had to survive in harsh environments since their inception. So, as our bee-searchers Hunt and Chittka put it,[2] 'The pervasiveness of such false memories generates a puzzle: in the face of selection pressure for accuracy of memory, how could such systematic failures have persisted over evolutionary time? It is possible that memory errors are an inevitable by-product of our adaptive memories.' So, arguably, the bees muddling up which flowers actually provided them with nectar was a *good* thing, in the sense that the ability to muddle up memories is the by-product of a brain that can change, learn and reason. Occasional memory mistakes are but a small price to pay for that.

This adaptive quality of our brain is called neuronal plasticity, and it is only due to neuronal plasticity that we can have any memories at all. The cells in our brain, called neurons, connect with one another to develop meaningful networks, and these networks change in accordance with new experiences. If we were unable to incorporate new information into our existing neural networks, we would be unable to change our thinking or behaviour in light of new evidence, and we would be hard-pressed to deal with any alteration to our environment. It is also through this process that we are able to record and incorporate both the positive and negative experiences we have with others, and this can ultimately help us differentiate friend from foe.

Every time we have an experience we can potentially form a memory of it, a memory that exists in the brain as a network of neurons. This could be a semantic memory of a fact, such as that Obama was the president of the US in 2015. It could be an autobiographical memory of, say, the time you went to London to see a show. Or, it could be a memory of a decision-making process such as how you solved a puzzle. For any kind of experience to stick around in the form of a memory, it needs to form a physical representation in your brain.

Today we know a great deal more about how this happens because technological advances such as fMRI have allowed us to take photographs from inside of the brain, allowing us for the first time in human history to directly see what live memories look like. These advances have enabled researchers to study the biological and chemical mechanisms underlying memory processes, and to test purely physiological theories of memory formation. We now know far more about memory than even just a decade ago, and are able to map memories from their birth to their decay.

Memory stamping

The process through which an experience is laid down as a physical memory representation in the brain is known as biological stamping. In order to stamp new experiences into long-term memory, a biochemical synthesis is required to make connections between the existing neurons in our brain.

Our neurons have thin arms known as dendrites that allow them to physically stretch out to other cells; spines on them act as the communication centres between these arms. Within each individual neuron, messages move mostly as electrical impulses, but neurons mostly communicate with each other through chemicals passed via synapses. A synapse is a gap (or cleft) between two neurons. Synapses are transmitters and receivers; strong memories are largely the result of a continued easy flow of information from one cell to the next. This communication happens through chemical messengers called neurotransmitters that tell neurons what to do, most notably whether they should become more or less active. We could think of the neurons as two airports with planes (neurotransmitters) flying between them. Depending on the runways (receptors) available on the synapse they reach, some of these planes will be able to land and some will not. This controls the flow of information between neurons, making sure that we are not burning out our neurons when we are highly stimulated.

I remember one of my professors demonstrating the concept of synapses and their associated cells in a memorable, and adorable, way. He stood in the middle of the lecture hall filled with about 200 students and waited patiently until he had our attention.

'I am a neuron,' he proclaimed, matter-of-factly. He whipped out his arms to stand like the letter T. 'These are my dendrites.' Then he opened his hands, flexing his fingers, which had been fists until this point. 'These are my spines.' He called up another student and asked them to stand beside him in exactly the same pose. He brought his fingertips to his neighbour's, creating a tiny space between them, 'And these are my synapses.' Finally, he took hold of his neighbour's hand and shook it, representing the way impulses could travel through his neuron body into the body of his neighbour.

Our brain has approximately 86 billion neurons already in place, so recording a memory is largely an act of making and adjusting connections between existing brain cells, rather than making new ones. While all parts of neuronal connections can be modified, most researchers argue that it is primarily the synapses that are important for memory formation.

Long-term potentiation is the process through which connections between neurons are increased due to a strengthening of synapses. This strengthening happens because the neurons are strongly or repeatedly activated in relation to each other. For example, you might be on a beach in Spain for the first time in years, feeling really relaxed. This will activate the neurons in the 'beach' network, along with those in your neural networks for 'Spain' and 'relaxed'. If an experience activates these connections strongly enough, or if similar experiences do so repeatedly, a long-lasting connection between these networks will be created; an associative memory linking the concepts of 'Spain', 'beach' and 'relaxed', for example.

One of the most prominent researchers in this area, who made fundamental strides in our understanding of what memories actually *are* from a biochemical perspective, is Michel Baudry.[3] In 2011, he and his team at the University of Southern California published a review of over 25 years of their work, in which they essentially boiled the biochemistry of memory down to two things: a process called long-term potentiation and the influence of a class of substances called calpains, calcium-dependent proteases. Baudry and his team say that calcium is needed to stimulate the proteins in our brains that allow our synapses to undergo memory-related changes that can last. When a connection between neurons is repeatedly or strongly activated, like an association between memories ('park' and 'trees', say), calpains are activated at that exact location. The calpains then change the structure of our synapses, leading to a stronger connection between the activated memory cells in the brain. It seems only when calpains come for a visit that we see the transformation from simple experience to lasting memory.

Sea slugs and rat brains

Also studying this phenomenon is Eric Kandel. I have never met this amazing winner of the Nobel Prize in medicine, a pioneer in memory research. I have, however, research-stalked him for years, following his papers, textbooks, autobiography and interviews. And through this I feel as though I know him. Kandel started an obsession with sea slugs in 1962, and along with his colleagues and students at Columbia University in New York, he continues to do research on the sea slug *Aplysia*. *Aplysia* is a portmanteau of ancient Greek words meaning 'sea' and 'hare'. Apparently they called them sea hares because of the little protrusions on their squishy heads which resemble the ears of a hare.

Kandel chose the slugs as research subjects because they use a simple system of neurons to remember their experiences and react to them. For example, if you pinch a sea slug's gill in an experimental setting, it can learn to react by withdrawing it. The neurons involved can be isolated and extracted, and grow at an astonishing pace. Neurons can be kept alive away from their host brains *in vitro* if they are placed into a life-supporting oxygenated fluid.

Since the sole purpose of neurons is to make connections and form a brain, isolated neurons immediately begin to search for other neurons to network with. To do this they grow longer dendrites and more synapses. According to Kandel,[4] 'new growth of synapses occurs in front of your eyes over the course of a day'. This exceptionally fast growth, which is far faster than the growth of human neurons, makes the sea slugs ideal candidates for studying how memories are formed within and between individual cells. And because humans rely on almost all of the same neuronal processes as these invertebrates, the research has direct implications for human memory.

The sea slugs have taught us a tremendous number of things

over the past few decades, and have contributed much to what we know about memory today. One of the most recent findings, detailed in a series of papers published by Kandel's lab in 2015,[5] [6] is that one of the proteins responsible for long-term memory is unlike most other kinds of proteins – it is called a *prion.*

Prions, short for 'proteinaceous infectious particles', can change in shape, folding and reshaping in structurally distinct ways. Another of their notable properties is that they can either exist on their own, or they can form chains. These chains can automatically trigger neighbouring cells to join the chain and therein make a physical connection. Until their image update in 2015, many of us, if we had heard of them at all, would have associated prions only with negative diseases like Alzheimer's and BSE (mad cow disease). The reputation of prions was so bad that Kandel pre-empted people's potential bad reaction, saying 'Do you think God created prions just to kill?'[7] before going on to explain their crucial role in memory.

The primary role of prions in memory formation seems to be to stabilise the synapses that constitute long-term memories, thereby adding permanence to the physical changes that have already taken place due to long-term potentiation and the influx of calpains. Calpains are like the architects of the synapse, planning how communications should flow, while prions are the construction workers who make the changes more permanent.

But just because a connection is fixed now, this does not mean it will remain fixed forever. Calpains and prions can come back at any time and change things once more. In 2000, researchers Karim Nader, Glenn Schafe, and Joseph Le Doux[8] at New York University examined the issue of memory fragments that change on a purely biochemical level, conducting an experiment where rats were played a particular tone and then given an electric shock. When they were subsequently played the same tone, they would freeze in fear, indicating that a memory associating the tone with a painful shock had been successfully generated.

Because learning to fear a particular situation or place is inherently linked with an emotional response, the rat's fear memory was expected to be in the amygdala, a centrally located part of the brain which looks like two walnuts (one in each brain hemisphere) and which is largely responsible for emotion. In the next stage of the experiments, the researchers gave rats the same tone-followed-by-shock treatment, but injected a compound called anisomycin that inhibits the formation of proteins – proteins like calpains – directly into part of the amygdala after the rats received the shock. What they found was that these rats did not react with fear when they heard the tone again. In other words, the rats were unable to form new long-term memories of something that scared them because the compound had stopped the proteins in their brains from working properly, further emphasising the crucial role such proteins have in memory formation. This block has to be introduced quickly, however, as the biochemical process of stamping begins almost immediately during a learning exercise or personal experience.

That isn't the whole story. After either 1, or 14 weeks, the rats who had formed an association were played the tone again to cause activation of the fear memory, but without the shock. After either 1 week or 14 weeks, the rats who had formed an association were played the tone again but not shocked to cause activation of the fear memory. However, if they were *then* injected with the protein block, they subsequently stopped reacting to the tone exactly as if the original memory formation had been interrupted. In other words, the memory had been destroyed. Any time the rats were stimulated to recall the fear memory, it could be interrupted by the researchers if they introduced the compound.

What is more, the rats only forgot the association if the

compound was present during the memory recall. If they were given the memory-blocking drug anisomycin in isolation, without being played the tone to make them recall the memory of the shock, nothing happened. This indicates that the drug does not inherently just make any particular kind of memory dissolve. Rather, there seems to be an interaction between an *activated* memory in the brain and this drug that leads to the memory being erased. By introducing a protein synthesis inhibitor immediately after or during recall of a long-term memory, the reconsolidation of the memory is stopped; the memory stops being stored in the brain.

This leads us to one of today's most *en vogue* biochemical theories of memory: retrieval-induced forgetting. This theory states that whenever we remember we also forget. So, while it seems intuitively appealing that every time we recall a memory we consolidate it and form a stronger and more accurate memory, this is far from the truth. Instead, every time a memory is recalled it is effectively retrieved, examined, and then recreated from scratch to be stored again. It is the equivalent of keeping a file of index cards, pulling one out to read it, throwing it away, and then copying out a new version on a fresh card for filing once more. And this is thought to happen *every time* we recall *any* memory.

In 2013 researchers Jason Chan and Jessica LaPaglia[9] at Iowa State University explored this phenomenon in humans. They conducted a series of experiments demonstrating that every time an experience stored in long-term memory was recalled by participants this cycle of encoding and storage was repeated. They used no drugs in their studies, only interviewing. In one of these studies they showed participants a video of a fictional terrorist attack. They then asked them to recall what happened. After the participants recalled the memory they were then given misinformation – when they correctly recalled that the terrorist had used a hypodermic needle on a flight attendant, they were incorrectly told that the terrorist had used a stun gun.

When later asked to recall the event once more, participants recalled the incorrect information (the stun gun) and were unable to recall the actual event detail (the needle). The researchers claimed that this meant the new information had actually replaced the original memory. So if bad memory interviewing introduces inaccurate information, it can actually lead to a restructuring of the biochemical stamps of memories in the brain with non-medical procedures. This is how retrieval, if interrupted, can actually induce forgetting in a number of ways. It makes every event, every time it is recalled, physiologically vulnerable to distortion and forgetting.

Another way to inhibit memory in both rats and humans on a biochemical level is through a drug many people will already have heard of –commonly known by its trade name, Rohypnol.

Roofies

The notion of memory-altering drugs is a mainstay of our social consciousness. So-called 'roofies' in particular have taken centre stage, with their ability to have only little detrimental impact on our current state of mind, but temporarily destroying our ability to form new memories. Rohypnol – pharmaceutical name flunitrazepam – is a member of a class of drugs known as benzodiazepines. 'Benzos' are used recreationally due to their interactive effects with other drugs such as alcohol and heroin. In non-recreational settings, they are most commonly used for their anti-anxiety, anti-convulsant, muscle-relaxant and sleep-inducing properties – in emergency rooms benzos are commonly used to sedate patients. In criminal contexts, they are known as a type of date-rape drug.

In a nutshell, benzos are a type of depressant. Many of us have likely heard that alcohol is also a depressant. This brings to mind someone alone in a bar crying into their drink. In reality, a depressant has little to do with sadness. A depressant simply depresses, or slows down, your bodily functions. Rather than thinking *sad* when we hear depressant, we should think *slow*. Slow, like problematic walking. Slow, like slow reaction time. Slow, like falling asleep. More specifically in the case of benzos, they slow down the operation of central nervous system. This, in turn, shuts down our ability to form new memories because it affects the biochemistry of our brains.

What, more specifically, happens when we ingest benzos that causes amnesia? According to neuroscientist Daniel Beracochea from the University of Bordeaux in France,[10] benzodiazepines are particularly known as acquisition-impairing molecules. This means that they prevent the formation of the protein synthesis necessary for the biological stamping of memory, much like the drugs the rats in our earlier mentioned research were given. Benzos are also generally considered to produce only anterograde amnesia, not retrograde amnesia. This means that they do little to events encountered just before their administration, but can severely impair memory for experiences after they have been ingested.

For those of you who just cried 'Show me more biology!', your wish is my command. Looking at their effect in greater detail, it seems benzos enhance the effect of the neurotransmitter GABA (gamma-aminobutyric acid). According to a review article published in 2006 by Daniel Beracochea: 'Specifically, sedative and anterograde amnesic effects of benzodiazepines were mainly attributed to α_1-containing GABA-A receptor subtypes.' For those of us who are not medical doctors, what he is saying is that the impairment is due to changes in the sensitivity of parts of the synapse that respond to GABA. Once again, it seems that changing what's going on in the synapse changes our ability to ever even form memories.

In classic research settings, such as the memory studies conducted by French scientist Pierre Vidailhet since the 1990s,[11 12] the effects of benzos are typically investigated by giving participants the drug before asking them to complete a task such as memorising word lists or geometric patterns. Since benzos do not affect short-term memory, it is often hard to tell that a person is under their influence as they seem to think and act normally; however, if they are tested after a time delay, they are unable to recall the items on the lists, sometimes even forgetting that they were given lists to recall at all. Similarly, if we are given a benzo in a hospital setting before an operation, we will likely forget any conversations we had with nurses, doctors and loved ones right before, during, and right after the surgery.

I experienced this first-hand when I was put to sleep for a hospital procedure, which was then completed more quickly than anticipated. As expected, afterwards I was conscious, talking, and coherent, but clearly unable to form memories. My partner, who had come with me, apparently asked me the same ridiculous questions every few minutes, to see whether I realised that he was just on repeat. I did not. I answered as if for the first time, every time. I also apparently kept thinking I had just woken up, and that *now* I was fine. This is an interesting little side effect of not being able to remember the recent past, and some patients with severe amnesia due to injury report the same thing. My partner even gave me a notepad to write down my responses to his questions, and kept flipping the page so I could not see that I had written exactly the same thing a few minutes earlier. Of course, I cannot actually remember any of this, but he did show me the notepad as corroborating evidence.

Clearly, then, the brain chemistry which is so crucial for our ability to form memories can be manipulated by the ingestion of certain kinds of drugs. But memories are more than just biochemistry. Memories are networks.

Laser beams

Moving up from the tiny elements of biochemistry, we are able to talk about memory structures that we can more readily see through the use of neuroimaging techniques such as fMRI, and which can be activated in living beings. Here we are looking at neurons themselves, and the physical connections between them.

When you experience something various parts of your brain light up, which is to say that they are activated by a tiny electrical or biochemical charge passing through them. Those same neurons then remain responsible for their particular components of your memory for that same event. For example, for a single event you may have neurons in the visual cortex responsible for keeping information about what you saw, some in the auditory lobe keeping information about what you heard, and a couple in your somatosensory cortex keeping information about what you felt. The brain is therefore faced with the difficult task of finding connections between the neurons that you lit up when you had the experience, rather than reassigning neurons that are conveniently clustered together to store the memory. This means that in order to study complex memory, we must acknowledge and study these networks of neurons.

Scientist Gaetan de Lavilléon at ESPCI University in Paris and his colleagues took a unique approach to studying these networks. They wanted to see if it was possible to play with the protein structures that underlie our memories and to thereby change neuronal connections *in vivo*. Publishing their findings in 2015[13] in the journal *Nature Neuroscience*, they described an experiment in which they generated memories in ways previously not thought possible.

To do their experiment they opened up the skulls of mice and attached electrical wires very precisely to individual cells in the pleasure centres of the brain, and to an array of other areas. They wanted to create a link between so-called place cells, also known as grid cells, and pleasure. University College London neurobiologist John O'Keefe[14] actually won a Nobel Prize in 2014 for the discovery of these place cells, which act like an internal GPS, letting us map our environment and storing only this kind of spatial information.

O'Keefe and his colleagues left wires attached to the brains of the mice while they explored their environment, and noted exactly which cells were activated when the mice were in a particular location. Once they had identified these specific location cells, they could monitor them. When the mice later went to sleep, the experimenters then waited for the location cells to be spontaneously activated while the mice were dreaming. When they noticed that the mice were dreaming about the particular location they sent a jolt to the pleasure centre. This created an artificial memory, linking the place cells of a particular location with positive emotions.

The success of this technique was demonstrated by the behaviour of the mice. When the mice woke up, they chose to spend more time in their perceived happy place than anywhere else – even though nothing positive had actually happened there. This was seen as indicative that a false memory had been formed by changing the physical structures in the brains of the mice.

Similarly, Steve Ramirez, the late Xu Liu and their colleagues from MIT (Massachusetts Institute of Technology) wanted to see whether they could make artificial connections between memory fragments by shining laser beams into mouse brains. In their research published in 2013 in the journal *Science*,[15] they claim 'We created a false memory in mice by optogenetically manipulating memory engram-bearing cells in the hippocampus.' Optogenetics is the field of science where light is used to control neurons which have been genetically modified to be photosensitive. This is done by attaching a light-sensitive protein called channelrhodopsin-2 to neurons when they are activated. For example, we can have a mouse remember a location and then attach the protein to those particular place cells. After that point, the cells can be turned on or off by the use of blue light. It's a bit like attaching a flip switch to cells.

Ramirez and colleagues found that activating a small but precise ensemble of mouse neurons in this way could lead to reactivation of a memory. They were able to erroneously pair old memories with new situations, thereby generating false memories. These false memories involved mice who had previously learned to associate fear of pain with one environment having this memory activated in another environment. They now erroneously associated pain with an otherwise non-threatening environment, essentially the opposite of the sleeping mice being given erroneous associations of pleasure with a given environment.

They specifically targeted cells in a part of the brain known as the hippocampus, a term I generally remember by thinking of a *hippo on campus*. Apparently this analogy is not quite right, though, because the word hippocampus comes from the Greek name for a large mythical seahorse; the structure is so named because it is shaped a little like a seahorse. The hippocampus lies pretty much in the middle of our brain and is responsible for our ability to navigate spaces, as well as for forming long-term memories. Note, however, that I get annoyed when people claim that memories are stored *in* the hippocampus, because this is a gross oversimplification – as we've just seen, memories are stored as networks throughout the brain.

The role of the hippocampus is more one of a mediator. According to neuroscientist Dean Burnett,[16] 'Information is channelled to the hippocampus, the brain region crucial for the formation of new memories and one of the only places in the brain where brand new neurons are regularly generated. The hippocampus links all of the relevant information together and encodes it into a new memory by forming new synapses. It's basically like someone knitting a terrifyingly complex tapestry in real time.'

Using optogenetics to steer how our hippocampus makes memories brings to mind the idea of *Matrix*- or *Total Recall*-type science fiction scenarios, where entire complex memories are directly implanted into people's brains using technology. We're not quite there yet but science is making quick strides regarding the technology we can use in the brain. Optogenetics really only moved from science fiction to science fact in 2010.

Then, towards the end of 2015 we found ourselves at the beginning of the sonogenetics[17] revolution, too. Sonogenetics involves changing cells using only sound – ultrasound, to be specific. It is perhaps too early to say where this technology could lead, but all kinds of exciting developments, along with justified ethical concerns about use and abuse, are no doubt on the horizon. From a futurology perspective, it may be possible to one day shine a light or use ultrasound on a specific part of someone's brain to alter specific memories and thereby give them a new personal history.

In order to contextualise all of this, it is important to emphasise the fundamental principle of memory – association is everything. It is the association between the individual memory fragments in different parts of the brain that makes what we think of as a whole memory.

I associate, therefore I remember

Association has been seen as a core characteristic of the mind since the earliest philosophers started trying to understand how we tick. The so-called laws of association were based on a conceptualisation set in place by Plato, and were formally written as laws in 300 BC by Aristotle, as the principles that he thought underlay all learning – learning, of course, being a process of memory.

According to Aristotle's *On Memory and Reminiscence*, there are four laws of association. The first is the law of *similarity* – the experience or recall of one object will elicit the recall of things similar to that object. Second is the law of *contrast* – the experience or recall of one object will elicit the recall of opposite things. Third, the law of *contiguity* – the experience or recall of one object will elicit the recall of things that were originally experienced along with that object. Fourth, the law of *frequency* – the more frequently two things are experienced together, the more likely it will be that the experience or recall of one will stimulate the recall of the other. We can still see these laws reflected in many of today's conceptualisations of memory, including those which we will explore in this chapter.

For 2,000 years these four laws were assumed to be true, but their importance was largely trivialised. At least this was the case until they were revitalised by John Locke in the 17th century, and then by Hermann Ebbinghaus in the late 19th century. Ebbinghaus was a pioneer during his time, being one of the first people to study higher cognitive functions in an experimental manner. He came up with a new way to study the development of memories, by training and testing himself on so-called nonsense syllables. Nonsense syllables are collections of letters which have no inherent meaning, such as OOB or KOJ. Ebbinghaus's idea was that these would be easy to memorise but would have no previous associations. In other words they were chosen because they should not skew results by having any existing meaning, as pre-existing meaning would make some of them easier to remember. While this has been challenged since, with researchers arguing that even nonsense syllables can still be assigned meanings, the effort was laudable.

In 1885, Ebbinghaus summarised his findings and published his magnum opus *Über das Gedächtnis*,[18] later translated into English as *Memory: A Contribution to Experimental Psychology*.[19] His experiments, which he conducted using himself as the only participant, gave us many insights into memory formation and storage that are still accepted widely.

The modern concept of associative activation is an elaboration on these original propositions by Aristotle and Ebbinghaus, and refers to the idea that there is increased activity in particular memories when other, conceptually similar, ideas or experiences are processed. For example, if you think about swimming, you will almost certainly automatically activate memories of the associated concepts of water, pool and bathing suit. This notion assumes that individuals develop a set of frequently used words and concepts. Each individual concept or word in the brain can be called a *node*. These nodes can be linked to one another to create complex ideas.

Nodes that have similar meanings are thought to have stronger connections. So, the node 'police officer' is likely to be very strongly associated with the node 'law' and very weakly associated with the node 'table'. Once we activate a node, the energy we send out to it can be conceptualised as radiating out from that original source and then activating other related nodes automatically. So if the node we activate first is police officer, the energy will then automatically travel out from there to other strongly associated nodes such as law.

This kind of associative activation is something that I think about every time I travel on the Underground in London, or the Metro in Paris. The network of lines could be seen to represent the brain, with each station representing a node. Just as we can get from any station to any other, so it is possible to move through a series of associations from any node in the brain to any other. Just as some stations are only a few minutes' direct travel from one another, while to get between others we need to go through a long and complicated series of connections and changes, so some associations between nodes are far stronger and easier than others, meaning that those associated memories come more easily to us. To extend the analogy even further, sometimes rail connections fail or break down, so we end up at a station other than the one we intended. In a similar way memory errors can be envisaged as occurring due to a physiological breakdown in the connections between nodes.

Associative activation can contribute to false memory formation at two points in time – during encoding and during recall. During encoding, it is possible to present a number of concepts to an individual, without ever mentioning the main concept. So, a researcher could mention the concepts law, man and uniform, without ever mentioning the concept police officer. However, because this concept was activated automatically due to its association with the others, it might also be encoded by the participant along with the others – they might think that a police officer was directly mentioned by the researcher when this was not the case.

A similar kind of mistake can be made during recall of an event; when trying to remember which concepts were engaged with earlier, an individual may remember that the concepts law and uniform were mentioned, and the sense of familiarity with the concept police officer (again, because it is automatically activated) may encourage the individual to incorporate this into their recall.

Associative activation thus implies that false memories are the downside of being able to form powerful associations. The upside is that these associations allow us to have memories in the first place, along with the ability to creatively rewire ideas to respond to our environment and come up with complex solutions to problems. This also means that if associations between memories or concepts can be strengthened or weakened, this can affect the likelihood of memory illusions and errors.

Who invited Kevin?

This perhaps all seems a bit abstract at this point, so let us try a more lively explanation. Everyone loves a party. The physical representation of a memory in the brain is commonly known as an engram. And, this engram needs to be able to connect with other physical representations, other engrams – memory is inherently associative in nature, so all of these physical representations need to be linked to enable us to form and access our memories. This means that every time you access a memory, you are essentially having an engram party in your brain.

Picture the scene: Engram arrives at a party. He is pleased to see that his two best friends are already there. Engram sits down and quickly reconnects with them. He has associated with these friends often and they have an incredibly strong bond – so strong that their connection feels automatic. Engram's besties are the concepts and ideas that are inherently and strongly linked to the specific piece of information Engram embodies. Let's imagine Engram is the memory of your favourite park. Those 'friendly' engrams, the most closely related concepts, are perhaps the location of your favourite pond and what the trees look like.

The next to arrive at the party are Engram's other friends and colleagues. Some of them he only has a weak connection with, and some of them he finds incredibly boring, but they usually still come to his parties. These are the concepts that still have connections with the memory you are activating, but they have weaker links. Perhaps these are the memories of a bench in the park, an adjacent street to the park, and the closest café. The pieces of information may not always be useful, and the associations are not necessarily strongly reinforced, but they are mostly still activated whenever the park memory is activated.

Engram spots Kevin. Who the hell invited Kevin? One of his colleagues admits that he let news of the party slip, and then felt it necessary to invite him. Nobody likes Kevin. Kevin seems to take it upon himself to ruin every party. Kevin is an unwanted memory fragment. Perhaps today Kevin is the memory of a news article you read on a recent park stabbing. Or perhaps Kevin is the emotional memory of something really embarrassing that you did in a park years ago. You do not really want to think about such horrible associations, but because they are activated automatically, there is very little you can do to stop yourself. Kevin isn't just unwanted, he is also very hard to shake off.

Luckily for Engram, just as he is fuming about Kevin's unwanted appearance, a friend arrives at Engram's party with whom he has not connected in an incredibly long time. Engram is excited to reconnect. They find that they still have many things in common, and their bond strengthens. This friend is the memory fragment that is activated through strange or atypical connections in content. Perhaps on just one occasion when you are thinking about the park you are reminded of a trip to Brazil that you took years ago. You are reminded of the lush parks in Brazil, and by increasingly thinking about this experience, you may strengthen the association you have between your local park and your Brazil trip. Engram has so much fun with this friend that he decides he should come to all future parties – you have successfully connected your Brazil memory with your park memory, so every time you think about the park in the future you are also likely to think of Brazil.

Two people from an adjacent table start talking to one of Engram's friends and they end up joining the group's conversation. One of them is incredibly boring. He has nothing exciting, unusual or memorable to contribute, and no one really pays attention to him. He eventually leaves, and it is likely Engram will completely forget he ever interacted with him. This is a situation when there is *potential* for a new association to be formed, but because of a lack of characteristics that are important for memory formation, the memory is not integrated in the existing network of memories in the brain. This is a situation where no memory is formed.

In contrast, everyone gets along swimmingly with the other new arrival. She has a lot in common with Engram and a number of other existing group members – she is so interesting and exciting that a few people even add her on Facebook. Here, a new experience has been stored as a new memory, and associations with relevant existing memories formed, which is stored in the brain and linked with existing structures. Perhaps you meet an attractive stranger in the park. The memory of that meeting becomes a new addition to the memories you recall when you think about the park.

The party is in full swing and some of Engram's friends are clearly looking for love. The attractive new addition is getting a lot of unwanted attention, and Engram also catches two of his friends getting it on in the spare room. Like people, memory fragments actively look for other memory fragments to make new connections. They are a bit promiscuous – they are willing to hook up with pretty much any other memory fragment if the situation is right. And, thank goodness for that, as it is these automatic attachments of engrams to one another that enable interesting ideas to form. Through our brains experimentally combining memories and ideas in novel ways, we get new associations – it is this that forms the foundations for our ability to be creative and artistic, to birth new ideas and solve complex problems. However, this same tendency can also lead to memory illusions when engrams become connected in ways that are inaccurate.

All right, party's over. Beat it, Kevin.

Fuzzy traces

One theory that attempts to explain more precisely why we can develop inaccurate memories builds on the theory of associative activation. It is called fuzzy trace theory. I have long thought this to be the cutest name for a theory, and I like to think that the mini-marathon of YouTube cat videos which I tend to use to introduce it in lectures gives my students a great way to remember it. It's an elegant theory that can be used to explain a lot of memory phenomena. It posits that remembering involves two things: gist and verbatim memory traces. Put simply, a gist trace is a memory of the meaning of an experience, and a verbatim trace is a memory of specific details. Most memories contain both gist and verbatim components. It might be helpful to think about a conversation between two people. Individuals are likely to remember both what a conversation was generally about – the gist – and exact words or sentences that were said – verbatim.

Memory scientists Charles Brainerd and Valerie Reyna at the University of Arizona[20] believe this theory can explain a diverse set of false memory phenomena and can be summarised as a number of major principles. Of these, I think four are sufficient to understand the fundamental mechanisms which enable memory illusions.

Principle 1: Parallel Processing and Storage. The first principle is that there is a parallel processing of inputs – individuals store 'verbatim' and 'gist' memory traces at the same time, and store them as independent pieces in the brain. So when looking at a scene we process what it looks like (verbatim) and what meaning or interpretation we assign it (gist) at the same time, and these two sets of information are stored separately.

Principle 2: Separate Recall. The second principle is that these gist and verbatim traces are also recalled separately. This means that one type of memory trace from an experience can be stronger than another. It also means that one, both, or neither type of memory trace might be accessible in response to a given situation. This helps explain why sometimes we may be able to remember someone's name (verbatim trace) but cannot remember what they are like (gist trace), and other times we may remember what a person is like but cannot remember their name. In the worst-case we remember neither; in the best-case scenario we remember both. The important point is that the recall of verbatim and gist memory can happen independently of one another, with gist memory being generally more stable over time than verbatim memory.

Principle 3: Error-proneness. The independent recall of the two types of memory traces opens individuals up to a host of potential memory errors. The inherently imprecise nature of gist memory fragments allows for feelings of familiarity derived from them about a given event to cause the fabrication of verbatim details. For example, an individual might recollect a gist trace that they had a conversation with their friend over coffee (gist), and this could lead to their erroneously placing the conversation at a specific local coffee shop (verbatim). This is a normal process in which an individual tries to make sense of their gist memories in a manner that fits with their personal history. Alternatively, an individual might have a strong verbatim memory of talking to their friend in a specific coffee shop, remembering the exact seats they took, and what they were wearing, but forgetting the gist of why they were there. Building on the strong verbatim trace, the individual may extrapolate and generate a false memory about why they were there.

While these illusions can happen spontaneously in everyday life, researchers are also able to generate them by intentionally misleading an individual's connections between gist and verbatim memory fragments. This is something that will be explored in detail in further chapters, as many false memory experiments use this technique.

Principle 4: Vividness. The fourth principle is that both verbatim and gist processing cause vivid remembering. When verbatim traces are recalled, individuals often seem to re-experience the items and specific contexts. Gist trace retrieval, on the other hand, is sometimes considered a more generic remembering, and is associated with experienced familiarity and the perception that something occurred but cannot be explicitly recalled.

As already mentioned, when gist traces are particularly strong, they can encourage what are referred to as phantom recollective experiences, which take the familiarity of the gist as a good cue for verbatim interpretations. An individual could feel a sense of familiarity when asked whether a friend partook in an event which they might plausibly have attended – I *feel like* he was there – and from this might generate an erroneous verbatim memory of actually seeing the friend at the event – he *was* there. In other words, the realism of gist and verbatim memory traces can be sustained when these traces are recalled separately from each other and recombined with others in such a way as to create a false memory.

These four principles give fuzzy trace theory a broad explanatory framework that can encompass many of the mechanisms proposed by researchers to help explain when, how and why false memories are generated. To summarise, fuzzy trace theory proposes that memory illusions are possible because each of our experiences is stored as multiple fragments, and these fragments can be recombined in ways that never actually happened. Clearly, our brains are biological and chemical marvels which have some built-in mechanisms on a physiological level that can lead to the generation of complex, physiologically-based, memory illusions. These potential routes for error are largely incidental as a result of the benefits of an associative memory system, for without these associations we would be unable to have the creative, adaptive minds we cherish.

4. MEMORY WIZARDS

HSAMs, braincams, and islands of genius

Why no one has infallible memory

How much information can the human mind store? This is a question that everyone asks themselves at least once in their lives. Perhaps while Googling frantically for ways to help remember the information we suddenly need to squeeze into our heads for an exam, a job interview, or something similarly daunting. Maybe there are tricks that can help me? Maybe these memory athletes know a secret that I don't? Or is it just genetics? *Oh, I hope it's not just genetics.*

Leaving aside worries about our own memories for a moment, we know there are individuals who can perform amazing feats of memory. They are sometimes called memory wizards: people who can verifiably recall important information at will – minutes, days or years later – in astonishing detail.

As described in their 2006 paper 'A case of unusual autobiographical remembering',[1] James McGaugh and his colleagues at the University of California researched this amazing phenomenon and were contacted via email by a woman they now refer to as AJ:

Dear Dr. McGaugh,
As I sit here trying to figure out where to begin explaining why I am writing you and your colleague I just hope somehow you can help me. I am thirty-four years old and since I was eleven I have had this unbelievable ability to recall my past, but not just recollections. My first memories are of being a toddler in the crib (circa 1967) however I can take a date, between 1974 and today, and tell you what day it falls on, what I was doing that day and if anything of great importance (i.e.: The Challenger Explosion, Tuesday, January 28, 1986) occurred on that day I can describe that to you as well.

I do not look at calendars beforehand and I do not read twenty-four years of my journals either. Whenever I see a date flash on the television (or anywhere else for that matter) I automatically go back to that day and remember where I was, what I was doing, what day it fell on and on and on and on and on. It is non-stop, uncontrollable and totally exhausting.

Some people call me the human calendar while others run out of the room in complete fear but the one reaction I get from everyone who eventually finds out about this 'gift' is total amazement. Then they start throwing dates at me to try to stump me. I haven't been stumped yet. Most have called it a gift but I call it a burden. I run my entire life through my head every day and it drives me crazy!

McGaugh and his team agreed to meet with AJ, although they were sceptical about what they would find. It's not unusual for people to claim to have superior memory, but when tested most of them fall far short of their claims. However, AJ agreed to be tested and probed with a wide variety of memory performance tests and she seemed to be different from all the other cases. The researchers give verbatim examples of her amazing abilities in their paper:

> *April 3, 1980?* 'I see it. Spring break. Passover, I went to that week. I was on Spring Break. I see the week. I was in 9th grade. The week before I was on Spring Break. I was into General Hospital.'
>
> *July 1, 1986?* 'I see it all, that day, that month, that summer. Tuesday. Went with (friend's name) to (restaurant name).'
>
> *October 3, 1987?* 'That was a Saturday. Hung out at the apartment all weekend, wearing a sling – hurt my elbow.'

Luckily for the researchers, AJ had kept a journal from the ages of 10 to 34, which they could use to verify many of her memories. The team believed AJ's ability was unique, and found it so astonishing that they thought it should have its own term: hyperthymesia (constructed from the Greek words *thymesis*, which means remembering, and *hyper*, meaning more than normal). Based on AJ's case, they asserted that hyperthymesics could be defined by two features. The first is that they spend a considerable amount of time thinking about their personal past; the second that they have an extraordinary capacity to recall specific events from that personal past.

This is different from superior memory, which is an enhanced ability to acquire and recall new *non*-autobiographical information. While AJ has an exceptional memory for her own life events she does not have the same incredible ability to recall information unrelated to her life, whereas superior memory individuals generally excel at this, being extremely good at retaining numbers and facts. The current view of superior rememberers is that their skills are the result of the application of memory strategies acquired through practice, and not innate abilities.[2] On the other hand AJ, whose memory abilities were exclusively autobiographical, claimed to be unable to consciously apply strategies to help her learn and retain other kinds of information. This was evidenced by her lack of success in school and bouts of unemployment – indeed she had few of the features one might tend to associate with such an amazing mind. Her memory for numbers and facts was nowhere near as good as her memory for her own life. So what can we learn from such cases about the limits of human memory? Is their perfect memory actually as flawless as it seems? And is such a 'superpower' actually desirable?

HSAMs

In the last few years hyperthymesics have largely come to be referred to as HSAMs; highly superior autobiographical memory individuals. The whole subject has only really gathered any scientific momentum since AJ came forward in 2005. According to research published in 2013 by Lawrence Patihis and his team at the University of California,[3] 'Highly superior autobiographical memory (HSAM; also known as hyperthymesia) individuals can remember the day of the week a date fell on and details of what happened that day from every day of their life since mid-childhood. For details that can be verified, HSAM individuals are correct 97 per cent of the time.'

After the research into AJ's case was published, more than 200 people claiming to have similar abilities contacted McGaugh and his team. There was a new sense of excitement in the scientific community – maybe this kind of ability was more prevalent than had previously been believed; maybe they just hadn't been looking in the right places. This influx of new claimants had the potential to create a whole new world of groundbreaking memory research. But, just as in many previous instances, case after case turned out to not meet the criteria of true hyperthymesia. The people who contacted them had good memories, but they were not exceptional like AJ.

Then, just as the researchers were about to give up hope, something magical happened. A second memory unicorn appeared: Brad Williams.

And, then another, Rick Baron.

Followed by Bob Petrella.

And in 2010 the group was even joined by a celebrity, the actress Marilu Henner.

According to reports,[4] at least 56 HSAMs have been identified around the world to date. It remains a small and exclusive club, but it's a great deal more useful and significant that having just one, solitary, case. Of course, the question on everyone's mind once they had found a whole herd of these amazing HSAMs, was, 'How does it work?'

While it is far too early to have a strong and well-supported scientific explanation for the phenomenon, there are a number of hypotheses.

Braincams

Perhaps memory is like a video recorder, keeping track of everything we do, and HSAMs are just better at using the playback feature than the rest of us. A landmark publication from 1952 simply entitled *Memory Mechanisms*[5], written by the American-Canadian neurosurgeon Wilder Penfield, suggests there may be some evidence to support this idea. One of Penfield's main research interests was treating patients with epilepsy by cutting through parts of their brain. While he had the brain exposed during surgery he also took the opportunity to poke around a bit. He would use electrical probes to stimulate different brain regions and ask his still-conscious patients to report what they experienced. Through this process he was able to identify the parts of the brain responsible for various senses and those responsible for the movement of various body parts. These sensory and motor cortices were mapped out so well by his technique that we actually still use his map today – it is referred to as the cortical homunculus.

What he found was that when he was stimulating certain parts of the brain, particularly the temporal lobes, his patients reported complex hallucinatory experiences (the temporal lobe is a large part of the brain which sits essentially behind your ears on both sides of your head). When electricity passed along the right and left hemispheres of the temporal lobe, patients reported hearing the voices of loved ones, or spontaneously hearing songs play. This region seemed to directly stimulate their auditory memories.

Penfield also worked with a colleague, Phanor Perot, to look separately at stimulation that invoked complex visual experiences. In research published in 1963,[6] they found that when part of the brain (the parietotemporal cortex) was activated, patients seemed to relive entire scenes from their lives. This effect, however, only happened when the right hemisphere was activated, which was taken to mean that perhaps visual memories were stored primarily on the right side of our brains. These recollections could often be confirmed as real memories, since they mapped onto documented experiences or everyday encounters in these individuals' lives.

The team thought they had found the neural substrate of past experience – the singular location in the brain that stored all our memories. According to Brenda Milner,[7] one of Penfield's former colleagues, this led Penfield to 'postulate that somewhere in the brain of each of us there is a continuous, ongoing record of the stream of consciousness (of everything we attend to, not of things we are not attending to) from birth to death'. A built-in tape recorder. A tiny webcam in our brains that is always on.

When questioned by Milner about this notion, which she had difficulties accepting as an experimental psychologist, Penfield apparently replied: 'Of course this is not memory as you psychologists understand the term when you refer to the variability of memory, with its abstractions, generalizations and distortions. In ordinary remembering we do not have direct access to the record of past experience in our brain.' So his idea was that not only did we have a tiny 24/7 camera that recorded this stream of consciousness, but also that this braincam stored its files in a secret location.

In his early work, Penfield speculated that this hypothetical storehouse of our memories might be in the temporal lobes; he later thought it might reside in the higher brainstem, then finally settled back on the temporal lobes. Like many other researchers at the time, he suspected that the hippocampus acted as the intermediary, allowing us to access, or not access, certain memories from this stream of consciousness storehouse. He therefore called the hippocampus the 'key of access' and sketched it in personal correspondence to Brenda Milner in December 1973. He claimed to have come up with the idea when he was studying a patient with postoperative memory loss. According to Milner:

> In his diagram Penfield was saying that the two hippocampi must play a crucial role in scanning, or retrieving, from that hypothetical record of the stream of consciousness . . . he was suggesting that if you are trying, for example, to remember something about John Jones, who was your friend between 1950 and 1960, then in some way, via the

> interpretive cortex of the temporal lobes, the hippocampi give you 'keys of access' to those past recorded experiences.

If Penfield was right, then everything we remember is stored away somewhere in our brains, and the abilities of HSAMs could be explained as their somehow having a more direct access to this information.

However, in spite of its intuitive appeal, the tape-recorder model of memory has lost scientific traction. Today the consensus is that there is no evidence of any such secret storage facility; there is no conspiracy whereby the hippocampus is restricting access to your full memory potential. In search of another explanation of exceptional memory, some people turn to yet another related concept: photographic memory.

Photographic memory

So, maybe we don't have a secret tiny braincam that records everything, but what about the ability to perfectly snap up individual moments? Perhaps we are able to make perfect engrams of, say, a particular scene. That seems like not too much to ask. A life-selfie. This one is going into the memory bank. *Snap*. A way to capture special moments in our lives. The scientific term that comes closest to describing the popular conception of photographic memory is *eidetic memory*, and the researchers who study it explore the limits of our ability to remember visual information.

The most popular way to study 'eidetikers', the term for people with eidetic memory, is through something called the picture elicitation method. Professor Alan Searleman from St Lawrence University, has described the testing process:[8] an unfamiliar picture is shown to participants on an easel for 30 seconds – researchers often call that 'unlimited viewing time', because most people neither continue encoding details nor care to after 30 seconds looking at the same picture. Try it – 30 seconds feels *really long* when you are stuck with a single image.

After the image has been removed the person is instructed to continue looking at the easel that it was on. They are then asked to describe everything they can about the picture. People who have eidetic memory will report that they can still see the picture, that they can scan and examine their personal memory of the image as if it were still in front of them. They generally use the present tense when describing a recently removed photo and can report a tremendous amount of detail.

Of course, most of us do not have eidetic memory, and never will. Searleman has summarised the research in one of the leading textbooks in this field, *Memory from a Broader Perspective*,[9] and suggests that eidetic images are different from other kinds of visual images people may encounter. The experience is not simply an after-image – the kind of lasting visual stimulation that happens when we stare at something for too long, or when a bright flash goes off and taints our vision temporarily. Those are the result of the way the cells in our eyes were stimulated. After-images are purely responses by these cells. After-images move with your eyes, and they are the opposite colour or shade to original stimulation. The aftermath of a white flash might be a dark circle right in the middle of your vision; the aftermath of red might be a light green. An eidetic image is different from this – it does not move with your eyes and remains the same colour as the original image.

Eidetic images are also different from regular visual memories which can arguably last forever; they fade away involuntarily and can last only a couple of minutes. Apparently the images usually fade away piece by piece rather than as a whole, and the eidetiker has no control over which components remain in place the longest.

Eidetic images represent a moment of amazing memory that cannot last. But, while much better than other kinds of visual memories, even they are still prone to manipulations, omissions and false inclusions – the same kinds of distortions as any other kind of memory. According to Searleman, even eidetikers can misremember entire objects and forget pieces of scenes; it seems their exceptional memories for a particular image can still have some flaws.

What is more, as far as we can tell this kind of memory really only exists in children. In one of the few reviews of the literature on this topic, dated all the way back to 1975, researchers Cynthia Gray and Kent Gummerman[10] estimated that 5 per cent of children have eidetic memory, and 0 per cent of adults do. It seems eidetic memory is quite possibly non-existent in adults.[11]

The rate of incidence is higher among children with developmental disabilities, particularly brain injuries, jumping to 15 per cent. These kinds of findings are what have led some researchers, such as Enrol Giray and colleagues,[12] to wonder whether eidetic memory is actually an immature version of memory, one that is used before we are able to think about and encode experiences in a more abstract manner. This could also mean that a child's eidetic memory is actually a sign of developmental problems, rather than developmental advantages. So, the ability closest to the popular conception of photographic memory is not actually all that impressive or common. It really is quite amazing that although by the late 1970s researchers essentially considered the existence of photographic or eidetic memory a myth (except as an incredibly short-lived and rare occurrence in children), the idea is still such a frequent misconception in society today.

So, how does this link back to our HSAMs? Well, on proper analysis, it doesn't. It might be tempting to draw a link to our HSAMs, but their memories are actually the result of a different process entirely. HSAMs are adults who may certainly *appear* to have photographic memories – their memories seem to remain in perfect, detailed, multisensory condition. However, they seem to experience their memory differently from eidetikers – not as a vivid and temporary visual echo, but as something more long-term. In 2013, Lawrence Patihis and his team from the University of California published the results of research[13] in which they had set out to examine just how perfect the memories of HSAMs actually are. Or, rather, how error-prone they are. The team asked 'Are people with HSAM abilities vulnerable to the same kinds of distortions and errors that others are, or do their abilities protect them in some way from suggestive influences?' To attempt to answer this question, they conducted three false memory experiments on 20 verified HSAMs.

In the first experiment, the HSAMs were given what is called the DRM[14] (Deese–Roediger–McDermott) test. In this test, participants are read a list of related words and asked to remember as many of them as possible. For example, they might be told to remember night, dream, pillow and dark. When they are tested later, most participants will recall other items that are conceptually related but which were not actually mentioned in the list; so if given the list just mentioned they might, for example, also include the word sleep. When asked about any of the false words they have recalled, participants often claim to be certain that it was originally mentioned, making this inaccurate word recall a tiny false memory. It turns out that in this task, HSAMs were as likely to misremember the non-mentioned words as control participants.

The second part of the study involved a classic 'misinformation' task. In this task, participants were shown two slideshows, each consisting of 50 photos. The first included a sequence of photos showing a man pretending to help a woman while actually stealing her wallet. The second showed a man breaking into a car and stealing money and necklaces. Later, the participants were given two narratives about the slide shows, each 50 sentences long. In each passage, six pieces of information were incorrect. So when the man had put his hands in his jacket pockets, the text claimed he had put his hands in the pockets of his trousers, for example. The researchers wanted to see whether the participants would report these inaccurate details when their memories of the slide-show events were tested. Contrary to what we might expect, HSAMs were actually significantly *more* likely to incorporate such incorrect details into their descriptions than non-HSAMs.

Finally, in the third experiment, HSAMs were questioned about whether they had seen particular footage of one of the 9/11 plane crashes, specifically United Airlines Flight 93, which crashed in a field in Pennsylvania. Here is an excerpt from one HSAM's quite detailed response:

Interviewer: . . . a witness on the ground in Pennsylvania took some video of the plane crashing and it has been widely shown on TV news and the internet in the months and years since the attack. Do you remember seeing that footage?

HSAM: Yes, but a couple of days later.

Interviewer: OK, Can you tell me what you remember about the footage?

HSAM: Uh, I saw it going down. I didn't see all of it. I saw, uh, a lot of it going down, uh, on air.

Interviewer: OK, do you remember how long the video is?

HSAM: Just a few seconds. It wasn't long. It just seemed like something was falling out of the sky. It was probably was really fast, but I was just, you know, kind of stunned by watching it you know, go down.

Interviewer: OK, so here is the last question, I would like for you tell me how well you can remember having seen the video on the scale from 1 to 10, where 1 means no memory at all and 10 means a very clear memory?

HSAM: I'd say about 7.

Of course, as this was a false memory study, the video was invented by the researchers and never actually existed – this type of study is referred to as the non-existent news paradigm, and interestingly HSAMs and non-HSAMs scored more or less equally poorly on it, with 20 per cent of HSAMs and 29 per cent of non-HSAMs reporting that they had seen the video and offering at least one or two details. It appears our memory unicorns may not be so magical after all; not even the best rememberers on the planet can have an absolutely perfect memory.

Picking up the memory-braincam analogy from earlier, in the 21st century digital photos, like our memories, can be quite easily shared and edited by ourselves and others. Even if photographic memory were to exist – and it seems doubtful that it truly does – it appears that memory Photoshop would exist right beside it. Along with advances in actual photography, our memory-photo analogy may need a similar upgrade. We no longer use Polaroid cameras, and we should no longer be calling memories *photographic*, at least not in a way that implies perfection and permanence.

Spreading activation

However, perhaps we can still find a way to explain why HSAMs may innately have far better autobiographical memory than other people, while still being prone to false memories. The most popular current explanation, much in line with the associationist model of memory that was discussed in Chapter 4, is that HSAMs have particularly strong connections between pieces of their memories, making for strong neural networks. More specifically, it seems that an enhanced version of something called the spreading activation model, as proposed by memory researchers Allan Collins and Elizabeth Loftus in 1975,[15] may best explain the process that happens in the brains of HSAMs. What this model suggests is that when we are looking for something in our brains, we send out an electrical signal searching for an idea, a piece of knowledge or an event that is related to whatever it is we are currently thinking about.

Let's say that we wanted to recall an event from our childhood. We might already be thinking 'I want to remember the family cottage.' Then, from the location in the brain of that initial piece of memory, the concept of your family cottage, we send out an electrical signal which disperses along the web of all those neurons connected to the initial memory. The memory engram of the family cottage will be connected to other engrams. As the neuroscientist Dean Buonomano at the University of California Los Angeles explains[16], 'An item is stored in relation to other items, and its meaning is derived from the items to which it is associated.'

It might help to understand the process if you picture a giant spider's web, with your initial search going out from the middle. In our hypothetical web, the things that are most strongly related to the concept are along the first, closely knit, threads, and they will be activated strongly and quickly. Then, as the signal progresses further from the centre, concepts further out – not as strongly related – can be triggered too. Of course in your actual brain the proximity of neurons to one another has little to do with how strongly connected they are, but a web of dissipating strength can help us to picture what is going on.

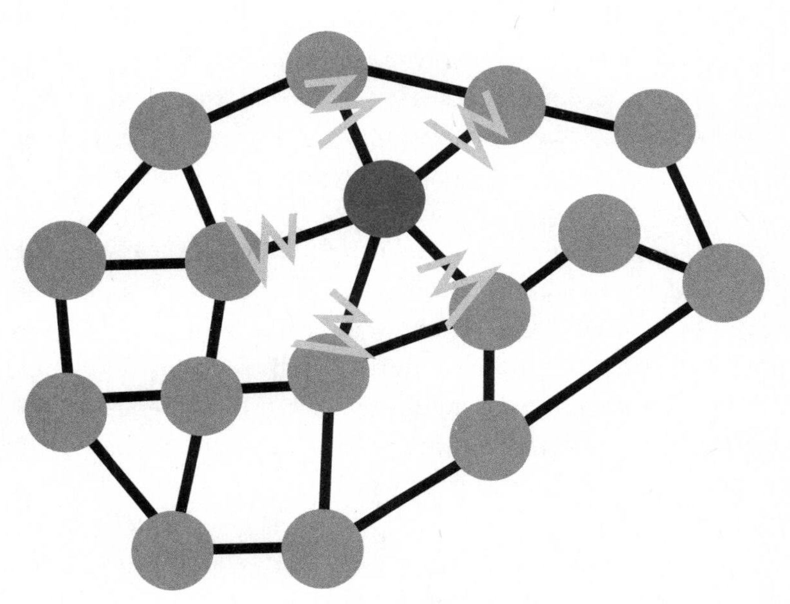

Spreading activation

So we might go from the family cottage engram, to the nearby lake, to the related concept of kayaking, to the yet physically further and more weakly related concept of islands, which then may make us spiral off and think about the unrelated concept of the Cayman Islands. The further out we go on our hypothetical web, the weaker the connection with the original concept, and the more likely we are to encounter topics that are not relevant to our original search. If this happens we can back up and restart our search from a more closely related topic. The process actually seems very similar to mind mapping, the activity sometimes used in educational and occupational settings where you write down ideas and concepts as an interconnected web spreading out from a main concept.

Let's return to this idea as it relates to HSAMs. AJ has previously described the subjective experience of her recall of events as a kind of cascade of memories that trigger one another. Fortunately, or unfortunately, her memories trigger one another automatically, rather than intentionally. She claims she is powerless to control this reaction, so she often activates too many memories at once. According to her, 'It's like a split screen; I'll be talking to someone and seeing something else.'[17] In the early stages of research on the issue, which is where we find ourselves right now, it seems that HSAMs may just have far more efficient, but not perfect, spreading activation in their brains. This could result in their being able to quickly search their memory banks and find the information they need in a way that is faster and more accurate than the rest of us.

Island of genius

Now let us explore another subject that is commonly associated with exceptional memory – autism. I bet right there you thought, yes, *Rain Man*. The movie *Rain Man* won an Oscar in 1988, and it depicts a grown man with autism who is very low-functioning, with severe social and cognitive deficits, but with an exceptional memory. A real person named Kim Peek originally inspired the movie.

So, would Rain Man be classified as an HSAM today? The answer is, probably not. He is certainly an individual with an innate ability to remember a tremendous amount of information, but his memory is largely for facts and numbers, rather than being autobiographical in nature (the A in HSAM). He is certainly, however, a superior memory individual. And he is not the only autistic individual to have such an amazing memory.

Approximately one in ten of those diagnosed with autism have exceptional skills related to memory. Such people are known as savants. According to psychiatrist Darold Treffert, who in 2009 summarised current understanding of the phenomenon,[18] individuals are referred to as savants if they have a serious developmental disability, due to conditions such as autism or brain lesions, but simultaneously excel at something, such as being exceptional at remembering certain kinds of facts. Their abilities range from splinter skills, where the individual has specific skills that stand out in relation to their disability (but which are not necessarily exceptional in relation to the general population), all the way to prodigious savants, who display such brilliance at something that it would be considered amazing even for an individual without cognitive deficiencies.

The particular island of genius a savant displays is clearly incongruous when we consider their mental capacities in general. For example, according to his mother Francis Peek, and author Lisa Hanson,[19] Kim Peek (the real Rain Man) had impairments in his ability to communicate socially, but had islands of genius expressed as an encyclopedic knowledge of music, geography, literature, history, and a number of other areas. The question, much like with HSAMs, becomes why and how individuals like Peek have such amazing memories.

This phenomenon of unusually enhanced memory is still referred to hyperthymesia when it occurs in conjunction with abnormal conditions such as brain damage or developmental difficulties. It is essentially the opposite of amnesia.

Autism has long been clearly linked to amnesia as well as hyperthymesia. In 1985 an autopsy on an autistic man by medical doctors at Massachusetts General Hospital, Margaret Bauman and Thomas

Kemper,[20] revealed the first glimpse that autism may be linked to abnormalities in the hippocampus, a part of the brain we have already mentioned as being associated with memory. Many researchers today continue to argue that this is the case; a team including medical researcher Simon Maier at the University of Freiburg, for example, found support for increased hippocampal size in individuals with autism spectrum disorders.[21] In a review of the science on this topic, neuroscientist Dorit Shalom from Ben Gurion University in Israel proposed in 2009[22] that in autism there is damage to part of the brain most often associated with episodic remembering of personal events, the limbic-prefrontal system, but other types of memory remain spared. This means that people with autism are going to more generally have worse memories of their own lives. This is different from what we would usually classify as amnesia, as it is not a complete lack of ability to form these memories, just a deficit.

Dorit also claims that high-functioning autistic individuals, who only have minor social disabilities but otherwise function normally, appear to rely particularly heavily on their semantic memories, while those with low-functioning autism rely mostly on their basic perceptual systems. What this means in practice is that those with high-functioning autism may tend to have a proclivity for remembering facts, while those who are low on the spectrum have difficulty remembering much of anything, leading to restricted and inflexible cognitive abilities.

It seems that the memories of savants are essentially the reverse of the memories of HSAMs. While HSAMs have incredibly enhanced memory of their own lives, but do not display any particular excellence in other kinds of memory, savants appear to lack autobiographical memory, but have an incredible memory for non-autobiographical things – facts and information. The memories of HSAMs are almost exclusively personal, while the memories of savants are almost exclusively *im*personal. They are the yin and yang of incredible memory. Maybe to have one type of extraordinary memory we cannot have interference from another type of exceptional memory? Or, perhaps due to limited cognitive resources, our brains simply cannot excel at remembering everything.

Further, what the autistic deficit in autobiographical memory means for the notion of the self and personal identify is also still a mystery. It seems intuitively likely that lacking autobiographical memory capacity may leave an individual perceiving themselves – and indeed other people – differently than the rest of us do. In support of this, individuals with autism are thought to have an underdeveloped theory of mind, or as Simon Baron-Cohen, Professor of Developmental Psychopathology at the University of Cambridge, has called it, mindblindness,[23] meaning the inability to understand the mental states of others and to appreciate that other people may have different emotions and desires from your own. Perhaps without the wealth of information from our own autobiographical memories to draw on it is harder to figure out not only our own selves, but also the selves of others.

Poison from the past

Because HSAMs and other amazing rememberers are such a newly identified group, we truthfully do not know yet why their memories work the way they do – we can only theorise. Savants are also extremely rare, and so the same is true of them. But one thing is clear: while you and I may see HSAMs as having an enviable superpower, for most people who actually possess it, having an exceptional memory appears to be more of a curse than a blessing. Just ask HSAM Alexandra Wolff, who said in an interview with

NPR,[24] 'It seems like you hold onto everything, and it seems like you're just stuck in the past all the time.'

Or HSAM Joey DeGrandis, who told *New York Magazine*,[25] 'I don't even know what it means when someone says, "I've let that go – it's out of sight, out of mind . . .". The other HSAMers I have met seem to share similar traits: the need for approval, seeking attention, putting themselves out there a little bit, maybe being a little sensitive to criticism and having issues with depression and closure. They are all contributing members of society and it doesn't seem like any of us are so hindered that we've ceased to function like a normal person, but there is a commonality in that we seem to be a little more sensitive and we sometimes have trouble with our emotions and we can be more prone to depression and it must be related to the fact that we remember in the way we do.'

Some researchers argue that the abilities of HSAMs are so rare because evolution has selected this feature out as it is actually an evolutionary disadvantage to remember everything.

You see, forgetting is generally considered to be important. According to neuroscientist Dr André Fenton at New York University, 'forgetting is probably one of the most important things that brains will do'.[26] Much like we need to be able to suppress distractors in our environment all the time – filtering out the conversations around us, the sights and sounds, other browser windows we have open, and so on – in order to focus on whatever task we may need to perform, we also need to avoid getting distracted by memories that are irrelevant to our current situation.

In a neuroimaging study led by Brice Kuhl at Stanford University, published in 2007,[27] researchers looked at the importance of the suppression of irrelevant memories. They had their participants study pairs of words that didn't naturally link, such as tomato–chips, bike–chair, water–night. Next, they showed participants one word and asked them what word it had been paired with. So if presented with tomato, the participants would ideally say chips; if presented with bike, ideally they would say chair. Crucially, the researchers would leave some word pairs out. Then, after this retrieval practice, the participants were tested on all the word pairs, while having their brains imaged by an fMRI machine.

The researchers found that the rehearsed pairs, those pairs that had come up during the practice session, corresponded with reductions in activity in the parts of the brain responsible for detecting and dealing with memory competition. In other words, the participants in their study needed to filter out less distracting information to find the word they needed. This is evidence that the more information we forget related to a concept, the stronger links between the remaining relevant information become.

As the researchers put it, 'These findings indicate that, although forgetting can be frustrating, memory might be adaptive because forgetting confers neural processing benefits.' We become more efficient rememberers if we filter out the less relevant information. It allows us to become better at remembering the important stuff in life. So, if we cannot filter out memories that are less important, as HSAMs seem to be, we may be at a disadvantage.

Another situation in which we can see the downside of a memory that is difficult to forget is in PTSD, post-traumatic stress disorder. One of the major symptoms of PTSD, according to the DSM-V (Diagnostic and Statistical Manual of Mental Disorders), the leading handbook for clinical psychologists, is intrusion: 'Recurrent and intrusive distressing recollections of the event, including images, thoughts, or perceptions.' Most people would agree that if something terrible has happened, we may not wish to have an exceptional memory of it, especially one that is hard to suppress.

According to a team of researchers led by psychiatrist Olivier Cottoncin from the University of Lille, France, people suffering from PTSD often experience a disturbance in the working of their autobiographical memory. They become overly fixed on a particular event. In a paper they published in 2006,[28] they supported this claim with an experiment involving 30 patients with PTSD, using another word-list paradigm. However, instead of using pairs like the Stanford fMRI study, they used what is known as a directed forgetting task. This is an experiment where participants are shown a series of words, and after each one is presented, are instructed to either remember or forget that particular word. Afterwards they are immediately asked to report as many of the words they were instructed to remember as possible, while omitting all those they were asked to forget.

In Cottoncin's study, those suffering from PTSD ended up remembering significantly fewer words overall, and fewer of the words they were instructed to remember, than individuals who did not have PTSD. They also remembered more of the irrelevant words. What the team inferred from these results was that those with PTSD were less able to forget irrelevant information, leading to problems remembering the items they were actually supposed to. PTSD sufferers showed an increased deficit in the inhibitory processes associated with memory compared to the rest of us; that is, they were less efficient rememberers because their brains were bogged down with parts of their past they were unable to forget. Causally, it is difficult to say whether these individuals had PTSD in the first place because they were already worse at forgetting things, or whether experiencing a trauma rewired their brains.

Overall, from looking at all the various cases of exceptional memory, one thing is certain: no one has an absolutely perfect memory. Indeed no one *can* have a perfect memory. But we should be grateful for that. We may not all have powerful recall for particular kinds of information, like HSAMs or savants, but when our memories are working at their best we remember most things pretty well. Functionally our memories turn out to be well rounded, able to deal with the many different kinds of information that are launched at us on a daily basis.

What is more, our memories are built to forget. Forgetting is a beautiful mechanism that trims down our neuronal connections to make our brains more efficient at storing only the information that is most important to us. When we realise the beauty of forgetting, we also see that perhaps the ability to remember *everything* would be a super-burden rather than a superpower.

5. SUBLIMINAL MEMORIES

Baby learning, psycho-phones, and brainwashing

Why we need to pay attention in order to form memories

Imagine a world where we could flick our smartphones to 'learn Spanish', go to sleep, and – *bam!* – wake up with a whole vocabulary course completed. Or, perhaps next time we want to completely forget an unpleasant event, just walk over to our local hypnotist and ask for the memory manipulation special to have the offending engrams removed. How about giving up smoking or achieving sustainable weight loss? No problem – just some music with backward messages stealthily embedded within it that says smoking and sweets are bad.

There are a lot of products on the market touting the effectiveness of hypnosis for helping us change bad habits, or of audiotapes that can 'reprogram' us while we sleep. However, in order to understand the very notion of such claims we must first understand the role of attention for memory.

On my first day in the first memory class I ever took at university I remember the professor picking up a piece of paper. He waited for the eager class of 150 students to settle down, then held up the unfolded sheet of paper and proclaimed: 'This is what happens in the world around us.' He then folded the paper in half. 'This is what you perceive.' He folded the paper in half again. 'This is what you pay attention to.' He folded the paper in half again. 'This is what you are interested in.' Another fold. 'This is what the brain makes into engrams. And this . . .' (he folded the paper one final time; it was now a fraction of its original size) 'is what you are able to access and recall later on.'

The confused class looked around at each other. What was his point? He broke the silence by saying 'Let's make sure this piece of paper is as big as possible when we are done with this class.' It was a clever opener. He had used the piece of paper as a memory aid to help the class remember some of the core components of the memory process itself. And right after the very fundamental principle that we need to perceive things in order to encode them – in other words, we need to see, hear, feel, smell or taste things – he had placed the importance of attention. Why did he do this? Because attention is a prerequisite for memory formation. Put simply, attention is the glue between reality and memory. If we do not pay attention to a stimulus in our environment, we cannot remember it. It's as simple as that. Attention and memory cannot operate without each other.

This basic principle is even true for things that we are actually looking at, or otherwise perceiving, but are not actually paying attention to. Think of those times in school when you were looking at the teacher but were thinking about something completely different and were unable to process – never mind later recall – what the teacher had actually said.

The importance of attention for memory even helps to explain why we are often bad at remembering people's names when we are first told them – because in the process of meeting someone new we are often dealing with so many new pieces of information at once that we do not pay sufficient attention when they say their name. We are busy attending to other things: *Nice tie. Where is his accent from? He seems nervous. I wonder if he is single. What is that cologne?* Everything but his name.

Okay, I just boldly declared that memory *needs* attention, and as a general rule of thumb that's true. However, if I'm being completely honest, there is a research study that appears to threaten this view. In 2012 a team of researchers led by neurobiologist Anat Arzi at the Weizmann Institute of Science in Israel[1] looked at memory during an inherently inattentive state – sleep. The results of their study, when published in the journal *Nature Neuroscience*, were given a straightforward title: 'Humans can learn new information during sleep'. The team found that participants who were exposed to certain smells accompanied by sounds during sleep formed associations between them, so that when exposed to the same tones during waking hours, they started to sniff as if searching non-consciously for smells. The participants did not have memories that they could intentionally recall of ever having been exposed to the smells, so when a particular tone was played they could not correctly specify the smell with which they associated it. The researchers argued that they had made very simple memories, as evidenced by the sniffing behaviour they displayed.

Somehow this basic association formation seemed to have overstepped the usual attention requirement for memory creation. Many researchers, in fields as diverse as child psychology, medicine, psychotherapy and cognitive science, have tried to achieve similar results. Some of these have found ways of understanding our attentional process that can help us to maximise our ability to remember information, and many have found themselves debunking common memory myths. So, what do scientists today understand the precise nature of the link between attention and memory to be? Let's start at the very beginning, by looking at attention in babies.

Baby Einstein

Everybody loves babies. At least it seems that way, as photos of friends and family with their tiny self-replicas certainly seem to pop up often enough. People share their babies' accomplishments as if it is the first time humanity has ever witnessed the development of an infant. Look at his first smile! His first drawing! His first word! Sometimes it seems that in this constant chatter of tiny baby updates, both in real life and online, everyone is trying to convey just one simple thing: 'Look at what a wonderful, astounding, unique creature my baby is!' And while it's easy to roll your eyes at it all, the truth is that babies *are* pretty amazing, already well on their way to utilising the intelligence we pride ourselves on as a species. But in order to do that they need to learn, and learning is inherently a memory process – to gain lasting effects from exposure to things in our environment we need to be able to remember them.

And if you're one of those parents who are anxious that their child should be brainier than its peers then you may be tempted to turn to products marketed specifically at the so-called diaper demographic, touting brain development and memory skills for children under two. People hear from a study published in 1993[2] that listening to Mozart improves college students' test scores, so they play Mozart to their babies to help them develop their problem-solving abilities. They find videos that claim to train their babies' intelligence, teaching them how to understand the world and equipping them with ways of thinking. They invest in sign language skills for babies. There's a whole world of baby learning, supposedly all based on solid science.

The companies that market baby media of this sort often provide testimonials from parents that show just how much their little Jim or Janet benefited from the experience. In these, the parents often mention the exceptional amount of attention that their infants pay to the videos. If attention equals memory, then this may be the key to their effectiveness, right?

Wrong. While attention is (apart from in exceptional circumstances) a prerequisite for memory formation, at the same time attention does not necessarily mean a memory will form. And when talking about attention in babies we often refer to just one thing – how long a baby looks at something, known by researchers as attentional gaze. As we know from our own lives, simply looking at something for a period of time by no means guarantees that we are giving it our full attention; we also need to be internally focused on the information that we want to encode and remember, we need to recognise patterns, and we need to be able to filter out all the other unimportant information that we may be taking in at the same time.

For infants, especially young infants, this is likely to be an impossible task. Babies do not yet have the ability to decipher which part of the magical glowing screen they have been placed in front of is most significant. They may not even find it the most attention-worthy thing in the room. This makes the chances of their managing to filter out and retain the take-home messages which their parents are so keen for them to absorb seem remote. Never mind that in the very early months a baby may not even be able to physically *see* the complex educational video they have been placed in front of, since their eyes are still unable to focus more than a few inches past their own noses. So, no, unfortunately your baby probably isn't creating useful memories from the TV while you do the dishes.

But don't just take my word for it. In a research report from 2010, 'Do babies learn from baby media?' Judy DeLoache and her team from the University of Virginia[3] studied how well 12- to 18-month-old children learned language from a popular brand of baby media. They found that children who viewed the educational videos for four weeks did not learn any more or any fewer words than if parents were given no instruction to teach their children language at all. However, they did find that the tots learned significantly more words if they were not exposed to any video but instead were taught words during everyday activities. It seems babies prefer the live show. Other studies have produced similar results. Live presentations of language, and of tasks, have been shown to be far more effective for developing babies' memories than any kind of media simulation.

Usually the effect of media on baby development has been found to be non-existent, but one large-scale study on the industry actually points to *negative* results – in 2007 Frederick Zimmerman[4] and his team from the University of Washington found baby television exposure to have highly detrimental effects on language development. They called 1,008 parents of young children and asked them about their children's media viewing habits. They also asked them to complete the short form of the MacArthur-Bates Communicative Development Inventory which measures language development in children. In this correlational exercise, for every hour of baby media watched per day by infants between 8 and 16 months, they were found to know six to eight fewer words.

The ineffectiveness of baby media is considered so well evidenced within the academic and professional communities that major paediatric bodies have provided clear-cut guidelines on the issue. For example, in 2011 the American Academy of Pediatrics (AAP)[5] clearly said that children under two should have no screen time at all, meaning no iPads, iPhones, laptops or TVs. Instead, parents should use play and live interaction if they want to give their babies the best possible developmental help. This may be bad news for the 90 per cent of parents who let their babies regularly interact with screens, and in practical terms may not be possible, but an effort can be made to limit screen exposure as much as possible. Next time you are looking after a young child and need to get on with some chores, perhaps choose crib time, grandma time, or playpen time, instead of screen time.

All this goes to show that attention works with a complex array of other physiological and psychological processes to enable memory formation. Simply looking at something and reacting to parts or all of it is insufficient. We may even ask ourselves whether that kind of interaction is a true representation of attention at all. But if that isn't attention, what *is*?

Happy to be blind

You are blind. And you should be happy about it. What's more, you are not alone. We are all blind. You see, scientists may disagree on some nuanced points about attention, but they all generally agree that it is the selection of some information for further processing and the simultaneous inhibition of other information. Paying attention to something requires you to be blind to the overwhelming majority of the information you are receiving from both your external and internal environment. It is your trusty filter that allows you to sift through the constant chatter of your senses and thoughts, the chatter that tries to tell you that you are hungry, that you are a bit cold, that the person sitting next to you is wearing a neon shirt, that you need to call your parents later, that your knee hurts, that there's an interesting conversation happening in front of you, that you like this song, and, oh yeah, how about getting some work done. Amazing human being that you are, you can usually handle it all without even noticing that you are doing it.

There are a number of experimental studies that have

demonstrated just how 'blind' we can be when paying attention to something. One of the most famous was conducted by Daniel Simons and Christopher Chabris at Harvard University and published in 1999.[6] They asked participants to watch a short video of a group of people passing a ball, and to count the number of times the ball was passed. After the video ended, participants were immediately asked to write down the number of passes, before being asked a series of unusual follow-up questions, which included 'Did you see a gorilla walk across the screen?' Of course, one of the most frequent responses was 'Why would I have seen a gorilla?' There was every reason for them to have seen a gorilla – during the video a woman in a full-body gorilla costume had walked right through the group of people passing the ball, at normal walking pace. In a phenomenal demonstration of our selective blindness when paying attention, 46 per cent of participants had failed to notice the gorilla due to their preoccupation with the task of counting the passes. This effect is called change blindness.

Change blindness can happen not only when we observe photos or videos, but also in real life. Psychological scientists Daniel Simons and Daniel Levin published a study in 1998[7] which showed us that when a stranger asks us for directions, if we are briefly distracted and the stranger swaps out with another person in the middle of the conversation, we are unlikely to notice that we are suddenly talking to someone else. Another psychological scientist, Ira Hyman, and his team from Western Washington University showed in 2010[8] that we can even fail to notice a unicyling clown if we pass them on the street while we are on our cell phones. Change blindness is the reason that in our everyday lives we miss our partner's haircuts, or say things like 'he came out of nowhere' when we are driving. The phenomenon is common or potentially even universal among animals, having recently been observed in both pigeons and chimps. It appears that even when we look sometimes we don't see.

Change blindness is a function of two bottlenecked processes, processes that need to filter a great deal of information and can only do so much at once. The first is our limited ability to perceive the world through our senses. The second is our limited short-term memory capacity. As mentioned in Chapter 1, our short-term memory really is super-short-term, lasting only about 30 seconds, and has a very limited capacity. That means that when we experience a complex scene we cannot possibly remember all of the details in it.

There may also be a third reason. Ira Hyman argues that we experience change blindness because we have conceptual representations of our experiences in our memories. These representations can be rather abstract – they are the gist memory traces we discussed in Chapter 3. Hyman explains: 'I have, for example, a rather vague picture of what my friends look like. Surprisingly, my interaction with my friends and the world is better because of this. I need to recognize my friends in different clothes, lighting, locations, and after they get a haircut.'[9] These memory functions that seem like failures may thus exist because they offer us larger adaptive advantages.

It appears that many of us are not just change blind, we are change double-blind. In 2000 Daniel Levin and his team at Kent State University[10] demonstrated that most of us engage in a metacognitive error called change blindness blindness. They asked participants to rate how likely it was that they would notice change in four different situations. Three of these situations had been previously tested and had produced change blindness rates in 100 per cent of participants; the fourth was the experiment just mentioned where participants were approached by a lost pedestrian asking for directions and the person switched during the conversation after being briefly hidden from view. However, across the four conditions, between 70 per cent and 97.6 per cent of participants thought they would detect the changes described, and they did so with high confidence ratings – we apparently grossly overestimate our live scene processing abilities and underestimate our own change blindness.

So it seems that our facility for paying attention overwhelmingly works to make us notice only a small amount of information so that we have a chance of actually processing it, and, in certain situations, remembering it for the future. Memory feeds into attention to tell it what 'important' information is, based on past experience, and attention feeds back into memory to update our internal representations of the world. But while researchers may not all agree exactly how this process does and does not work itself into lasting memories, they all agree that sleep is an inherently inattentive process. And yet, it sometimes feels like we can learn or remember things that happen while we sleep – as the experiment where sleeping participants were exposed to different scents while different musical tones were played seems to suggest. So what actually happens to our memories when we sleep?

Replay

I generally consider sleep an annoying necessity. If I could, I would skip my nightly time-out altogether. What makes sleep all the more frustrating is that scientists are not even entirely sure why we need it. We know how much we need, seven to nine hours. We know what it is like, split between being almost completely unconscious and hallucinating vividly. We also know under which circumstances it is most likely to occur: total quiet and cool darkness. But why do we need it? Simply for rest and replenishment seems like a decidedly insufficient answer, as we could presumably also gain those benefits by lounging on a couch without actually losing consciousness.

Biological psychologists Gordon Feld and Susanne Diekelmann at the University of Tübingen in Germany argue in a review paper from 2015[11] that dreaming is a state of 'active offline information processing essential to the appropriate functioning of learning and memory'. They also suggest that our memory engrams, and the connections between them, are played back to us during deep sleep, like a replay of the day. In particular they claim that something called 'active system consolidation theory' can help us understand the association between memory and sleep. This theory suggests that during a type of sleep called slow wave sleep, the memories that we just formed while we were awake are strengthened. This, they argue, is how sleep helps to consolidate memories: by repeating the connections between neurons, and replaying our experiences, which makes some of our memories last.

According to neuroscientist Gordon Wang and colleagues at Stanford University in 2011,[12] sleep particularly appears to be important for bringing down brain activity from the levels it reaches during the day, and perhaps diminishing some of the less important connections to increase brain efficiency – the previously mentioned process of synaptic pruning. Wang's team argues that this process allows us to keep our most important memory traces and get rid of the less important 'noise of daily experience'.

The other thing that seems to necessitate this down-regulation of the brain provided by sleep is the reliance of our brains on glutamate. Most of us know the word 'glutamate' from the food additive monosodium glutamate, MSG, and MSG is chemically related to the glutamate found in our brains. Glutamate is the most common neurotransmitter in the brain, and works to open up some of the main channels of communication between cells. These channels allow calcium to flow into cells, which activates them and allows the chemical encoding of engrams, enabling us to generate and access the networks of information required for complex memories. Thus, our brains release glutamate as part of the chemical process that underlies memory formation. This glutamate mainly remains in the brain until it is processed and drained when we sleep. But while we need glutamate to make memories, too much of it is bad for us. An excess of it can cause excitotoxicity, in which brain cells are damaged and killed due to an overactivation of glutamate receptors causing an excess build-up of calcium. It seems that sleep allows us to drive down our overall glutamate production, essentially preventing brain cell self-destruction.

However, not everything is necessarily down-regulated during sleep. In 2014 researchers including Guang Yang from the New York University School of Medicine found that catching some z's after learning increases the formation of synaptic spines,[13] which are considered one of the foundations of memory storage. Synaptic spines are tiny doorknob-shaped bumps on our dendrites (the connections between our brain cells), and they are where most of the synapses in the brain are located. Generally, by increasing synaptic spines, we improve our memory. Yang and team found this when they had mice learn a new motor skill, running on a fast rotating rod. They then looked into the brains of the mice, having injected them with a protein that made the relevant motor cells fluorescent yellow so their growth over time could be monitored. They found that mice that had been sleep-deprived after the learning task formed significantly fewer spines than those that were allowed to sleep, and thus had worse memories.

These memory processes that occur during sleep may even help us understand why we dream – we often dream about events, people, situations or emotions similar to those that transpired during the day. We know that memories are being variously pruned or reinforced as we sleep; in the process related memory engrams may be activated simply because of their associations, and it may be that they then manifest as dreams. Of course quite often dreams can be bizarre combinations of engrams that could never occur in reality.

So, sleep seems to be a way for us to strengthen, reorganise and transform memories. When it comes to consolidating new or complex memories, the old adage is right; it is indeed 'best to sleep on it'. But, could we tap into this dream state and learn *new* complex information?

Psycho-phone

The 1920s saw the invention of a device called the psycho-phone. It was formally patented in 1928 by Alois Saliger, a businessman from New York who, according to a 1933 interview in the *New Yorker*, was 'a tall, spare, thin-lipped man with piercing eyes and a wide forehead'.[14]

The device he had created was a record player triggered by a clock, so the device could activate itself once the owner was asleep. Once this 'time-controlled suggestion machine'[15] triggered it would begin to play a recording of Saliger speaking. He would talk in the first person, in a soothing voice, starting with a note about how the buyer was asleep and telling them that their subconscious would now follow his spoken guidance. Then he would begin the sleep therapy, repeating phrases such as 'Money wants me and comes to me. Business wants me and comes to me . . . I am rich. I am a success . . .'[16]. According to Saliger the device worked because 'it has been proven that natural sleep is identical with hypnotic sleep and that during natural sleep the unconscious mind is most receptive to suggestions.'[17]

A quick search online reveals that similar audio files aimed at helping the buyer achieve their dreams are still widely available, sold as 'sleep learning' or 'subliminal learning'. They make a wide range of promises, from 'develop extreme motivation', to 'overcome social anxiety', 'think yourself thin' or 'increase your memory by as much as 75%'. There are even anti-ageing programs on offer. According to a scientific review by health psychologist Madalina Sucala at Icahn School of Medicine in New York, along with her international team,[18] in 2013 there were at least 1,455 hypnosis smartphone apps offering, essentially, a high-tech version of the psycho-phone.

These sorts of products make tall claims with major potential applications. It is easy to see how the military, professional organisations and educational outlets would be immediately interested in the possibilities should they prove effective, as indeed they were during the last century. In order to test these claims of the possibility of subliminal learning more scientifically, in 1956 a series of studies was conducted by weapon researchers Charles Simon and William Emmons at RAND Corporation, a company that conducts research for the US Armed Forces.[19] Presumably they wanted to know whether this was something they could use in military training, or even something they could weaponise. They tested responses to material presented at various levels of wakefulness, and used electroencephalography (EEG) to confirm that their participants were actually asleep. In fact, while it seems like a basic prerequisite for such research, they were among the first scientists to ensure that their participants were actually sleeping during a sleep study.

Their conclusion: 'The results support the hypothesis that learning during sleep is unlikely.' They found that exposing participants to learning materials during sleep had no discernible effect. This led to such claims falling generally out of scientific favour.

Most researchers considered the issue a closed case, and thought that there was no need for further studies. But a few always remained hopeful, and research did continue, albeit at a very slow pace.

In a 1995 series of neuroimaging and behavioural experiments examining fear responses in rats, Elizabeth Hennevin and her team at the University of Paris[20] claimed that animals *could* form new associations during sleep, and that information to which an animal was exposed during sleep could have behavioural implications for when they awoke. In particular, they claimed these effects were achievable during a type of sleep known as paradoxical sleep, which exists alongside our slow wave deep sleep. Paradoxical sleep is characterised by rapid eye movements (REM) and by brainwave patterns similar to wakefulness – hence the paradox, the brain is acting as though awake, but the person is not. Hennevin and her colleagues went on to generalise that the same sleep-learning processes should take place in humans, since our brains are similar to those of rats in many important ways. If the brain is in an awake-like state then might the person be able to attend to stimuli, at least in a basic sense? If so, paradoxical sleep could indeed be the key to sleep learning.

Similar assertions about enhancement of learning during sleep were tested almost two decades later in 2014, utilising new knowledge gained from neuroscience and sleep research, by a team spearheaded by Maren Cordi from the University of Zurich in Switzerland.[21] The team wired up 16 adult volunteers to electrodes attached to devices that measured several physiological responses. They used EEG to measure brain activity, electromyography (EMG) to measure muscular activity, and electrooculography (EOG) to measure eye movements. Together these would give a strong indication of whether the participants were actually asleep, and whether they were experiencing deep sleep or paradoxical sleep.

Sessions started at 9pm; after being hooked up to the equipment participants were left to sleep from 10.30pm until about 2am. They were then woken and presented with a learning task that involved remembering the location of 15 pairs of cards displaying animals and everyday objects (similar to the card games sometimes known as 'memory' or 'concentration'). This learning task was paired with exposure to a particular smell. Then the participants were asked to go back to sleep, and when the physiological measures indicated they had entered paradoxical sleep, they were either exposed to the same smell as during the task or to no smell. Finally, after being awoken again, the subjects were tested on their memory of the word pairs in a room with no smells. The idea was that being exposed to the smell would reactivate the memory of the task and thus reinforce it.

So what did Cordi and her team find? Nothing that would support the hypothesis that we can learn, or even reinforce, new complex information while we sleep – they found no improvement in memory if participants were re-exposed to smells during paradoxical sleep.

Can we actually learn new complex information, or significantly reinforce memories, while we sleep, as the subliminal learning advocates suggest? The answer is a definitive *no*. There is no evidence that we can learn words or facts, or benefit from any sort of personality-pumping propaganda, while we sleep – not even when information is delivered in the convenient form of an iPhone app. As Madalina Sucala and her research team reported in 2013, 'technology has raced ahead of the supporting science'. The only people who get rich from 'sleep your way to success' subliminal learning recordings are the people selling them.

But are there other ways of influencing the non-attending mind?

The psycho-phone advocates were attempting to extrapolate and widely distribute access to a basic form of hypnosis. Perhaps going to sleep and then being hypnotised by an audio recording as a sort of afterthought is simply insufficient. Perhaps if such techniques are instead administered by a professional, starting when we are in a waking state, they could help us reap the kind of benefits sleep-learning products are striving for, maybe even letting us dig into the hidden vaults of our memories.

Hypnosis

'It's not a thing.' That is my standard response.

When my students, friends or family want to talk about hypnosis, I generally brush them off. Hypnosis is not a thing, *because I know it isn't.* I am arrogantly confident about the answer; dismissive, like the kind of person you would hate to have at a dinner party.

Yet my mom remains convinced she was hypnotised once, going onstage a sceptic and returning a convert. She is hardly alone in this – there are hypnotists who make their entire living from lavish demonstrations in which they are able to make people go as stiff as a board or take off their clothes. There are also hypnotists who claim to be able to use their skills beneficially in psychotherapeutic or medical contexts. A friend of mine who is a medical doctor insists that hypnosis is effective for pain reduction and can be used in lieu of anaesthesia. He insists that it works with as much vehemence as I insist that it doesn't exist.

More than a few people seem to strongly believe that hypnosis can help our memories, too. According to a 2014 review article by Guiliana Mazzoni of the University of Hull and colleagues,[22] one of the most persistent beliefs regarding hypnosis is that it can help us transcend our regular mnemonic abilities, enhancing our ability to remember new information and perhaps even allowing us to reach into our past to uncover buried memories

This sentiment is echoed by some very recent survey studies. In 2011, for example, a study by psychological scientists Dan Simons and Chris Chabris surveyed 1,500 adults in the US. They intentionally tried to sample as representatively as possible, to get a real feel for what the broader American population believes. Of those they surveyed, 55 per cent claimed that memory can be enhanced through hypnosis. In another study in 2014, also in the US, Lawrence Patihis and colleagues at the University of California, Irvine found that 44 per cent of university students believed that 'hypnosis can accurately retrieve memories that previously were not known to the person'. So perhaps I am being too harsh. All of these converts, with their strong beliefs and compelling stories of the power of hypnosis, might be onto something.

We can do a quick review of the academic literature and we find hits on a surprising number of research articles on the topic of hypnosis and memory. But not all research is created equal; the quality of these studies must also be considered. We cannot draw firm conclusions from weak research. But it seems the hypnosis researchers are aware of this as well – there is a set of guidelines for hypnosis research laid out by scientists Peter Sheehan and Campbell Perry, the earliest version dating back to 1976.[23] It states that 'no behaviour following hypnotic induction can be attributed to hypnosis unless the investigator first knows that the response in question is not likely to occur outside of hypnosis in the normal waking state'. In other words, we need to make sure that the effects we see from hypnosis are due to the technique itself, and cannot be due to influences that happen in normal everyday life. If your friend quite readily dances like a chicken if you ask him to when he is awake, then chicken-dancing is not going to convince scientists that your friend has been hypnotised.

Are we now ready to confront a slew of randomised control studies – often considered the holy grail of academic research – which vehemently disagree with some of our views about hypnosis? Most notably there are a number of recent meta-analyses demonstrating that hypnosis is effective for analgesic, painkilling, effects. It has support from medical researchers for use in minimising pain during surgery where no anaesthetic can be applied,[24] it can provide long-term symptom relief for some sufferers of irritable bowel syndrome (IBS),[25] and it can help people with fibromyalgia.[26] Even our friend smoking cessation makes an appearance, with researchers claiming that hypnosis can help you quit. However, research demonstrating the usefulness of hypnosis for helping to form new memories, or uncover old ones, remains curiously absent from the search results.

What it leaves me wondering is whether the debate I have with my mom and others is more semantic than I initially thought. This thing that at least some people call hypnosis does seem to be able to elicit significant results in some situations. It makes me wonder whether perhaps some people are using the word more loosely than I am, or in a different way.

We often refer to hypnosis as it was traditionally conceived. We think of it as an altered state of consciousness that can only be induced by a hypnotist; a special procedure that is somehow conceptually and empirically different from other non-hypnotic procedures; a procedure that is sometimes said to allow the participant to remember things they otherwise would not, sometimes even from their early childhood. I associate hypnosis with buzzwords such as 'trance', 'reconditioning' and 'unconscious barriers'. But perhaps this is my personal misconception, and not the definition that modern hypnotherapists themselves use.

Pretend for a second that you need to come up with a good definition of hypnosis. How do you begin? In 2011 there was a meeting of two of the major organisations of hypnotists in the UK – the British Society of Medical and Dental Hypnosis (BSMDH) and the British Society of Experimental and Clinical Hypnosis (BSECH) – that addressed exactly this question. It appears that I am not alone in thinking that the definition of hypnosis may need some clarification. According to the report that came out of this meeting, spearheaded by Irving Kirsch: 'The unanimous consensus was that conventional definitions of hypnosis and hypnotisability are logically inconsistent and that at least one of them needed to be changed.' Logically inconsistent?

To me, this inconsistency is just another reason why the concept of hypnosis is problematic. The report concluded that hypnosis is defined by suggestion to enter a hypnotic state. But it seems that the character trait of suggestibility, in this case known as hypnotisability, is the prerequisite to being hypnotised. It's a circular argument that makes it impossible to argue whether hypnosis makes people respond to suggestion, or whether suggestibility makes people respond to hypnosis. People who are hypnotisable, the report goes on to explain, are highly likely to follow suggestions made by another person regardless of whether or not they are actually 'hypnotised'. This is problematic as it takes us back to your chicken-dancing friend, who may indeed dance like a chicken regardless, making it impossible to study hypnosis independently.

Be you a clinical or dental hypnotist, when key definitions in your discipline not only clash, but actually contradict one another, you know you have a problem. Unfortunately the meeting failed to reach a consensus on the issue, so for the time being the contradictions remain, and formal definitions remain largely elusive. But this problem needs more than lip service. Definitions matter.

Not me

The thing is, even if we accept the existence of hypnosis in some vaguely defined sense, it turns out that many people cannot be hypnotised at all. Hypnosis researcher and professor of medicine David Spiegel from Stanford University[27] says that while the exact number of unhypnotisables is unknown, he estimates that it is about 25 per cent of us. Studies on hypnosis obviously need hypnotisable participants. Those beneficial effects reported in studies are evident only in samples that are hypnotisable, and many may even be limited to the *highly* hypnotisable, an even smaller number of people. Thus, if researchers say that 80 per cent of their sample enjoyed a particular benefit, it may sound great, but it may only apply to few of us in the real world.

If you are wondering whether you are one of the hypnotisables, look no further than the Stanford Hypnotic Susceptibility Scales. The Stanford Scales consist of a number of assessment activities to be undertaken under the guidance of a test administrator. As an example, a participant may be asked to hold out their arm. The test administrator may then suggest to them that they are holding a very heavy weight, telling them to picture the weight, to feel it pulling their arm down. If the person's arm starts to sink under the suggestion, the person is said to have passed that part of the test, indicating that they may be hypnotisable.

While this is how the test interprets such a response, other psychologists may just call this type of reaction a form of suggestibility or compliance, as both of these have to do with a person's willingness to follow instructions. Or, as the hypnosis researcher Graham Wagstaff[28] put it, 'there is a strong case for arguing that much of the special status that has been awarded to hypnosis may have resulted from a failure to consider the power of social pressures and the normal capacities of ordinary human beings'. He argues, as do many psychological scientists, that the positive effects we sometimes see when people have supposedly been hypnotised may result from regular phenomena such as relaxation, imagination and expectation.

So we're back to semantics. To those who simply wish to use hypnosis as a term intended to encompass those more regular psychological phenomena Wagstaff mentions, most notably the power of suggestion, I say go ahead. Telling people to close their eyes and listen to instructions, to picture wonderful things, or not to feel pain, may very well have some kind of beneficial result. Relaxing, soothing, positive, empowering – it all sounds great to me. It may even help us focus our attention on different parts of the brain, different perceptions, and different stimuli. It sounds like something that should – and seemingly does – have real, scientific, effects, but we can acknowledge this without having to resort to the mumbo jumbo of a hypnosis-specific state of consciousness.

People who are hypnotised are paying attention to what the hypnotist is saying; they are choosing to be hypnotised and to engage in the resulting behaviour. That means their attention is still engaged and functioning, allowing the possibility of behavioural and psychological consequences from stimuli they encounter. And while there is good scientific evidence that hypnotism can help with some medical and psychological issues, there is no such evidence to suggest that it has any kind of beneficial effect on memory. The notion that it does most probably stems from popular media – there are hundreds of books, TV shows and movies that portray hypnosis as a key that can allow access to hidden memories. Unfortunately, this is completely untrue.

If an event is suggested to a person who has been successfully placed into a susceptible state of the sort sometimes referred to as hypnotic, they are far more likely to imagine and generate false memories of impossible occurrences. In a study as far back as 1962, for example, medical scientist Theodore Barber[29] from Boston University found that many of those to whom it is suggested that they are being regressed to their early childhood display childlike behaviour and claim they relive their memories. When examined further, however, the responses given by these 'regressed' participants does not match what children would actually do, say, feel or perceive. Barber argues that it may feel to patients as if they are reliving their early years, but in actuality such experiences are creative re-enactments rather than rediscovered memories. Similarly, if used during therapy, suggestive and probing questions combined with hypnosis have the potential to generate complex and vivid false memories of trauma, something we will explore further in Chapter 9.

So, particularly as it applies to the world of memory science, my sentiment remains: Hypnosis? It's not a thing.

Brainwashed

In my own research I implant rich false memories of complex emotional events. I convince people they did things which they absolutely did not do, and they go on to tell me about those things in incredible detail. When I tell people about my work, they invariably ask me whether I use hypnosis to achieve this. After I explain to them that you do not need any such technique, they almost always follow up with, 'So how does your brainwashing work, then?'

To me, brainwashing sounds like something an evil villain in a Batman comic would do. The term was popularised in the 1950s, as populations awoke out of the devastating repercussions of two world wars and tried to explain what had happened; how it was that normal people could commit acts of such monstrous brutality as the Holocaust. In 1957 psychiatrist William Sargant[30] defined brainwashing as 'methods influencing the brain which are open to many agencies, some obviously good and some obviously very evil indeed . . . beliefs, whether good or bad, false or true, can be forcibly implanted in the human brain . . . people can be switched to arbitrary beliefs, altogether opposed to those previously held'.

In my personal conception of the term, brainwashing refers to changing a person's ideology or epistemology – changing their ideas about the world and the knowledge they believe they have of it. In certain kinds of false memory research, including my own, scientists have been able to have a small temporary effect on a person's view of the world – perhaps, for example, making them think they have committed a crime when they have not. I'll go into exactly how that can be achieved in Chapter 7. This arguably has the hallmarks of a brainwashing process but we take steps to ensure that no permanent skewing of a person's world view can occur by placing a great deal of emphasis on the debriefing. During the debriefing we ensure that the experimental process is rigorously explained, and ensure that any false beliefs that a participant may have acquired as a result of their experiences in the study are debunked. For me at least the term brainwashing also implies intent – the desire to reprogram people's fundamental ideologies – and this is absolutely not something my colleagues and I wish to do; we are simply interested in discovering how memory works.

While I would still be hesitant to call them 'brainwashing', preferring the more universal term 'influence', there certainly seem to be many examples of changing a person's thinking or behaviour without their knowledge in our everyday environments. Propaganda, the selective presentation of information to influence our views and behaviour, is all around us: buy this soft drink, it will make you happy and surrounded by friends; vote for this politician, they will make everything better; join the army, it will be exciting and fun. Certainly this kind of media is almost ubiquitous and has the potential to influence our everyday decisions. But in almost all of these situations, while we may not realise that this propaganda is changing our opinions, we do usually realise that we are seeing it. This has been referred to as *supraliminal* advertising, ads that we actively perceive. We know that we see billboards on the street and commercials on TV, and that things in stores are organised to grab our attention and money– we aren't idiots. These may have the same effects as brainwashing, but without being concealed and secret.

But, maybe there are things that we do not see that still influence us, *subliminal* messages. In 2012, researcher Dobromir Rahnev and his colleagues from Columbia University[31] argued that 'despite the general notion that attention and awareness are necessary for higher cognitive processing, recent studies are beginning to demonstrate that in some cases, complex behaviours can be influenced without conscious attention'. In support of this, a small study published in 2009 by Simon van Gaal and his team from the University of Amsterdam found that subliminal stop signs influenced participants completing a simple computer task that involved discriminating between two coloured circles. The stop signs in this experiment were presented so quickly, for 16.7 milliseconds, that participants did not report seeing them at all. In spite of this, the signs slowed down participant button-pressing in response to the coloured circles they were supposed to identify.

In another study on this topic, psychological scientists Mika Koivisto and Eveliina Rientamo from the University of Turku in Finland found that the same kind of effect can occur for a task identifying animals. They found that flashing a picture of an animal so quickly that the participant could not consciously perceive it made the participant faster at classifying another picture of that animal. For example, participants would be faster at deciding 'animal or not' for a picture of a horse if they previously saw a picture of a horse non-consciously. However, the researchers found that this effect only worked for this very basic level of 'animal or not', and did not improve participant reaction times for anything else, like different kinds of animals. These kinds of results indicate that this non-conscious process is likely to have incredibly limited ability to influence our behaviour.

While these kinds of subliminal effects are still poorly understood, and cannot always be replicated, they are likely due to an effect called priming. Priming is another memory phenomenon. It is a function of implicit memory, the process that allows our previous experiences to inform our present or future experiences despite us not consciously realising we are being influenced by those particular memories at the time. It is a form of memory that cannot be accessed in the way we normally think about memory – as a sort of visual, acoustic or tactile representation; this kind of memory is generally argued to be more primitive, closer to being an impression or feeling.

I *like* this idea. I *trust* this brand. I *feel like* I should slow down. This *seems* dangerous. Perhaps a smaller amount of attention, so small that it is not perceptible, is all we need to encode these deep-rooted feelings as memories. These are the kinds of friend-or-foe feelings that have allowed us as a species to make snap decisions that have enabled our survival over millennia. They are still memories, and they can have strong influences on us, but we cannot remember their origins.

When priming effects were first discovered, some people thought that this meant that any kind of subliminal stimulus would elicit effects, including things such as backward messages woven into songs (a process referred to as backmasking). The public feared that these kinds of effects could be used like brainwashing, for evil. Of course researchers also wanted to know whether messages played backwards could influence people. In a series of studies described in a paper from 1985 entitled 'Subliminal messages: Between the devil and the media', Jon Vokey and Don Read at the University of Lethbridge[32] set the issue to rest. They set themselves the question 'Is there any evidence to warrant assertions that such messages affect our behavior?' Their answer? 'Across a wide variety of tasks, we were unable to find any evidence to support such a claim.' According to them, and scientists who followed in their footsteps, we can neither process nor remember such backward messages, so we can rest assured that they can have no impact on our beliefs or behaviours.

So I hope this chapter has demonstrated that we need some form of attention to be able to create memories, and that sleep is crucial for the consolidation and strengthening of those memories. I also hope that it has shown that intelligence-enhancing baby videos, sleep learning, and hypnosis or subliminal messaging as ways to influence ourselves or others, are all best considered creative fictions.

6. DEFECTIVE DETECTIVE

Superiority, identity crisis, and making monsters

Why we are overconfident in our memory

As a criminal psychologist, when I prepare expert testimony I sometimes find myself siting at my desk reading case files where one thing after another seems to have gone horribly wrong. From problematic eyewitness testimony, to unreliable victim statements, to detectives who seem to misremember how evidence was collected, I find myself with a host of issues that warrant concern.

Conspiratorialists may argue that the police cannot be trusted, perhaps even suggesting that they may intentionally distort the facts of a case. I, however, choose to believe that it is not the integrity of the police that is problematic. I am confident that they almost always want to do their jobs as effectively as possible; they want to catch criminals and to protect the public. The problem is that they have been assigned an impossible task, where they need to piece together the past in a way that is entirely reliable. But, as we know, memory almost never is entirely reliable.

Unfortunately the police are often under-equipped to deal with the complexity of the memory problems that can plague investigations. In a study I published in 2015 with Chloe Chaplin[1] at London South Bank University, I investigated whether or not

British police officers knew more about memory and other psychological processes than members of the general public. We distributed a 50-item questionnaire and found that, overall, the police held as many misconceptions about issues in psychology and law as the general public, but that they were more confident in their responses. Of our confidently wrong police, 14 per cent endorsed the myth that 'Memory is like a video camera' and 18 per cent believed that 'People cannot have memories of things that never actually happened'. This research points to a lack of police education and potential problems with overconfidence, the latter being something we will explore throughout this chapter.

As much as we may wish our justice system to be infallible, and as much as we hope that the police always catch the right culprit, we know that in truth this is not always the case. There are plenty of instances of people being wrongfully convicted and imprisoned for horrific crimes. The Innocence Project[2], an organisation dedicated to getting innocent people exonerated through DNA testing, has helped to release at least 337 people who were wrongfully convicted. On average, these people served 14 years in prison for a crime they did not commit. Faulty memory played a role in at least 75 per cent of those cases. Those figures are just for the US, and just for cases in which DNA was available, so around the world there are significantly more people who have been wrongfully imprisoned.

When such cases are subsequently examined it often becomes clear that the police officers involved did everything in their power to get a suspect convicted. It might be easy to assume that the police have been terribly negligent or, worse, deliberately tried to frame someone they knew to be innocent. Perhaps sometimes that is the explanation, but it is also perfectly plausible that they simply got caught up in a string of psychological biases. The police can develop 'tunnel vision', where they overvalue evidence that supports their argument and discredit or ignore information that contradicts it.

And it's not just the police – this kind of process can happen to anyone, because incorrect information can seep into any of the coherent stories we construct to understand reality. To use a term stolen from one of the world's leading legal psychologists, Peter van Koppen, we can all be 'defective detectives', struggling to be unbiased evidence collectors.

As we will see in this chapter, when we need to make sense of an event, but do not have enough information to do so, we tend to import other plausible content to fill in the gaps. Events in our minds need to have a linear progression, connections, reasons. Once we have this kind of plausible narrative, we can become incredibly confident in its accuracy. But what exactly is the relationship between confidence and accuracy, and how does it all tie in with memory?

Above average

Let's switch gears for a second. Do you think you are a good driver? How about compared to your peers? As Ola Svenson from the University of Stockholm asked her participants in a 1981 study:[3] 'We would like to know about what you think about how safely you drive an automobile. All drivers are not equally safe drivers. . . . We want you to indicate your own estimated position in this experimental group.'

Rather than being interested in the self-assessed driving ability of participants, this was actually one of the first studies to look at overconfidence. It was found that the overwhelming majority of Americans and Swedes thought they were both safer and more skilful than the average driver. Svenson had even asked them to compare themselves specifically to the average in the particular study sample, which means that they thought they were better than peers of similar age and intellectual capacity.

People responding this way intuitively makes sense to me, since whenever I find myself on the road I cannot help but think that most of the other drivers are idiots. Similarly, other studies have shown that most people think they are more intelligent, more attractive and more competent than average. We may not necessarily believe we are absolutely brilliant at everything – far from it – but we do generally think that we are better than average at pretty much everything. Which is, of course, statistically impossible – if everyone thinks they are above average, clearly a lot of people are wrong. Yet, studies have found this overconfidence effect in all kinds of areas. Police are overconfident in their ability to detect liars. Students are overconfident about their course grades. CEOs are overconfident in their business decisions. Teachers are overconfident in their teaching ability. The problem is so persistent that in a 2011 article published in *Nature*, social scientists Dominic Johnson at the University of Edinburgh and James Fowler at the University of California[4] argued that 'Humans exhibit many psychological biases, but one of the most consistent, powerful, and widespread is overconfidence.'

One reason for this may be the superiority illusion, which suggests that we have a tendency to overestimate our positive qualities and to underestimate our negative traits. This is a characteristic that is inherently linked to memory, because in order to think about our positive traits we need to be able to remember the good things we have done in our lives that provide evidence of those traits. For example, you may think about all the times you have done chores around the house, and think to yourself that you are a really good spouse. You took out the trash, bought groceries, cooked, and did the dishes. Go you. However, you may be forgetting or diminishing the times when you did not do any of those things and actually made more work for your spouse, leaving them frustrated and with extra work to do.

In 2010 the web company Cozi conducted a survey of 700 men and women with children who were either married or in a committed relationship.[5] It aimed to find out how much each partner thought they contributed to household chores, and how much they thought their spouse contributed. This type of research is sometimes called a 'chore wars' survey. Perhaps unsurprisingly, women were thought to do more chores by both partners. But much more interesting than this finding was people's opinion on the comparative distribution of chores. If the percentages each partner thought they contributed to each chore were added together, they frequently came to over 100 per cent. Let's take, for example, the item 'scheduling of events and appointments'. On average dads claimed to do 50 per cent of the scheduling, and moms claimed to do 90 per cent. Of course, it's not possible to do more than 100 per cent of any given chore, so what is happening here? Perhaps the participants simply misunderstood how to assign percentages to these tasks, but I propose an alternative explanation: our memory is selfish.

We are less likely to remember something done by someone else than something we did ourselves. This is partly because watching a partner do a chore, or having them report to us that they did it, provides us with a far less rich and complex memory trace than if we had performed the task ourselves because there is simply less sensory input. The memory trace being weaker means that we are probably more likely to forget that it happened in the long run. On the other hand, we will always have stronger and more meaningful memories of those occasions when we have done chores. This means that unfortunately the game is always rigged against our partners – our recollections of our own contributions are always likely to be stronger and more significant.

Besides the superiority illusion, we also suffer from *survivorship bias*. This is an error whereby we tend to focus on successes and overlook failures, literally focusing on people or things that survived a process. This is the kind of mistake people make quite blatantly when they say things like 'Steve Jobs was a university dropout, so I'm going to drop out in order to make a success of myself.' By focusing on one success, they are failing to think about all the people they have never heard of who were in a similar situation and did not achieve fame and fortune.

In 2003, a study was published on survivorship bias amongst investment bankers. In it, hedge fund manager Gaurav Amin and Harry Kat from City University, London, looked at how hedge funds were invested between 1994 and 2001.[6] They found that many investments were ended early, and were omitted from a database that was used to calculate the risk of new investments. They argued that because generally it was bad investments that dropped out of this database it overly focused on investments that worked, leading to a survivorship bias. The authors argue that this bias means that 'hedge fund returns may be overestimated and risk may be underestimated'. Many economists have identified overly optimistic investing as a contributory factor in financial crashes, and such investing obviously arises from a basic bias as a result of only thinking about successes.

Similarly, a police officer may discount the times they elicited false confessions from suspects, at least partly because they may not even know that such failures happened – they have therefore omitted this information from their internal database of failures and successes, in a similar way to our investment bankers. If a suspect is in jail a case is generally considered closed – even if the person ended up there due to poor police practices. In such cases the police officer is likely to miss any failures on the part of themselves or their colleagues and instead count them as successes, falsely bolstering their self-appraisal. Lack of visibility of our own failures along with an excessive-focus on achievements leads to overconfidence in our abilities and assessment of opportunities, hence the survivorship bias.

There is one more illusion that may play into our tendency to be overconfident. It is related to the greater strength and accessibility of our memories of our own actions and insights compared to those of others – the illusion of asymmetric insight. In 2001 Emily Pronin at Stanford University and her colleagues published a paper[7] on this bizarre bias, appropriately entitled 'You don't know me, but I know you'.

Over six studies the team demonstrated that we think we know our close friends and roommates better than they know us. For example, in the first study participants were asked to think of a close friend and answer a number of questions about how well they knew them, including the extent to which they thought they understood their friend's feelings, thoughts, motivations and personality. Finally, they were asked whether they could see the 'essential nature' of their friend. This was done by telling the participants that we are all like icebergs, with part of our true selves observable by others and part hidden from view. The participants were then asked to pick a picture of an iceberg that best represented their friend from a selection showing icebergs at various levels of submersion. Then, participants did the reverse task, thinking about how their friend would answer these same questions about them. Five other similar studies were also conducted to examine this bias for different types of relationships, including for roommates and strangers.

Pronin and her team found that participants believed that their own quintessential qualities, including their intimate thoughts and feelings, were mostly kept internal but that those of others were more likely to be observable. They were more submerged icebergs, while other people were more visible icebergs. This makes sense from a memory perspective because we have direct access to our own thoughts and feelings and so appreciate that they can be complicated and nuanced – which makes them difficult for other people to understand. On the other hand, it can be difficult or even impossible to appreciate the complexity of the thoughts and feelings of others in anything other than a basic 'surface' way – we tend towards assuming that is all that there is to understand. Our general outlook is 'I'm a riddle, but my friend is an open book.'

This bias turns out to be really important for our decision-making and arguing skills. In their final study, Pronin and her team asked 80 participants to complete a background questionnaire on a number of politically relevant topics, including such items as whether they identified as liberal or conservative, or whether they were pro-life or pro-choice. Then, several weeks later, they asked questions about how well the participants thought their in-group knew their out-group, and vice versa. So, for example, they asked self-identified conservatives how much conservatives as a whole know about liberals, and they asked them how much they thought liberals knew about conservatives. They found that liberals and conservatives both claimed to know the other side better than the other side knew them, as did those on either side of the abortion debate.

Asymmetric insight helps explain why in arguments and debates we may believe that the other side will never understand our point of view. We may also think we perfectly understand their point of view, perhaps also bolstered by the superiority illusion that we are smarter and more informed than our opponents. As Pronin suggests at the end of her paper, we can begin to think 'I know everything about the other party, and I know they are wrong. They don't even try to understand my arguments. If only they knew more about it, they would be on my side.' It is an easy trap to fall into, and one that is a staple of political shouting matches.

So, overconfidence has far-reaching implications, from bias in our everyday internal dialogue when evaluating relationship fairness, to our inability to give our failures equal weight and acknowledgement to our successes, and our problematic assumptions about the knowledge other people have of us and we of them. It touches every aspect of our lives. Even if we wish to be humble and take pains to avoid overconfidence illusions, we may not be able to – they are largely the by-product of selective memory processes we cannot control.

These same processes make it easier to understand police officers responsible for putting innocent people in prison. Often they will have been overconfident in their ability to tell whether the suspects were guilty or not. That overconfidence may have been fuelled by several of the illusions we have described. The illusion that their police work was better than average, or the illusion that they only ever put guilty men in prison, or the illusion that their understanding of the case was deeper than other people's. It is these natural illusions, which can easily befall us all, that have helped contribute to many miscarriages of justice.

Forgetting we forget

While overconfidence can be the result of memory processes, memory processes can also be the victim of overconfidence. In other words, memory causes overconfidence which causes overconfidence in memory.

I'll try to make that a bit clearer. Prospective memory is our ability to remember to do things – we have already touched on it in previous chapters. It is the memory that is necessary in order for us to stick to goals and do things that are important for our future. It is like the personal assistant in our brains; our internal to-do list that reminds us that we have to go to the bank, go to the supermarket, clean the house, meet Sophie for lunch at 2pm, and so on.

It is an amazing ability. Yet, like all of our memory functions, it is far from perfect. How many times do we think to ourselves when we have an insight 'I'll remember that, I don't need to write that down', only to realise the next day that our brains are not as good at this task as our trusty Siri. Even when we have memory aids like phones or diaries at our instant disposal we overestimate our ability to remember information and may choose not to use them. This faulty appraisal process is why we forget meetings, forget to pick up packages at the post office, and on some days end up feeling like we're not really capable of getting much done at all. We are overconfident that we will remember, and we pay the price.

And we pay the price in more than one way. Marketing teams know this feature of our memory and they try to exploit it; companies seem increasingly keen to capitalise on our overconfidence. Many of them offer a free trial for a subscription-based service, get us signed up, and then have an automatic billing process for subsequent months. What they are doing is banking on us forgetting to unsubscribe before their fees kick in. And it works. Repeatedly. We keep falling into the trap, presumably because we think *this time* we will remember to unsubscribe.

In an article published in 2010, businessmen Jeff Holman and Farhan Zaidi[8] explore the economics of this prospective memory banking. They collected data on these kinds of free trials, and found that not only do many people stick with the services; they are far more likely to stick with them if the trial is longer. In their sample, retention was 28 per cent for those who were given a 3-day trial and a whopping 41 per cent for a 7-day trial group.

According Holman and Zaidi, 'Firms are increasing the length of their free trials, possibly to increase naive forgetting by consumers . . . [they] offer extended free or reduced-price trials – of lengths running in months rather than days or weeks – ostensibly to give their new customers as much time as possible to test and experience the benefits of their products, but effectively as a way to more fully capitalize on consumer forgetting.' Clearly not all those who apply memory science have our best interests at heart, and it seems we are consistently overconfident in our prospective memory abilities.

There is a distinction that should be made here between predicting future remembering and predicting future *changes* in remembering. Our subscription cancellation failure is a prospective memory example that applies to predicting future remembering – 'I will remember to do X.' Research has also been done into how we estimate future change in our memories – 'I will remember all the characteristics of X.' While predictions of future remembering are faulty, it seems predictions of future changes in memory are even worse.

Nate Kornell from Williams College explored this issue in research published in 2011.[9] He studied 430 participants in a memory-monitoring task geared around 'judgement of learning', where participants are asked to estimate how well they have learned something. It is generally assumed that people's estimate of their future ability to remember will be based on how strong their memory fragment of the information in question is after learning, so they will evaluate the strength of their memories, and evaluate stronger memories as being more likely to be recalled later. We go through this process every time we study for a test or presentation – we estimate how well we have learned particular information, and then use that as a basis to decide whether we should rehearse something again. It is a type of metamemory, a way of assessing our memory skills and predicting future remembering for a particular task.

In this particular study, Kornell asked his participants to study word pairs either once or four times. Then he asked them to estimate how well they would perform at a test administered either five minutes or one week later. When he then compared participants' actual and estimated performances, he found a clear stability bias. For example, for estimations regarding their performance one week later, participants guessed on average that they would be able to get 9.3 word pairs correct, but when they actually came back they only accurately remembered 1.4 word pairs. According to Kornell, 'People act as though their memories will remain stable in the future.' This kind of research, which has been repeated in many other learning contexts, shows that even though all of us know that we *can* forget things, we seem to systematically underpredict how much we *will* forget. To make matters worse, this effect seems to increase as time delays increase – in Kornell's study 'The results demonstrated long-term overconfidence: Relatively modest immediate overconfidence transformed into enormous overconfidence as the test delay increased.'

In another study demonstrating this phenomenon, published in 2004,[10] Asher Koriat and his colleagues at the University of Haifa in Israel showed that participants estimated that their memory would be essentially the same at immediate recall as *one year* later. So, we are already bad at estimating how much we will remember in the near future, and seem to be even worse at estimating how much we will remember in the distant future.

What can we do about our tendency to forget that we forget? Kornell has some direct advice for students: 'If today is Friday, and you feel ready for the test you have to take on Monday, don't take the weekend off. You might be right – maybe you are ready now. But that doesn't mean you'll still be ready by Monday. Indeed, you're likely to be way overconfident.'[11] And, for the rest of those everyday adult tasks we want to accomplish, 'Don't trust your memory. If someone asks you if you can remember something, say no. Write it down.'

Identity crisis

While some people are great at remembering faces, and others are great at remembering names, I am good at neither. If I ever meet you, I'm sorry in advance. I will probably introduce myself to you repeatedly on different occasions. This will almost certainly confuse you; we may have shared wonderful conversations and wine. I may even reference your own conversation or research back to you, without realising that you are its source. So, why is my memory for these kinds of interactions so terrible?

Well, there are individual differences in the ability to recognise faces. Not just in terms of remembering them, but being able to look at them and map their features – in the way that enables you to look at a photograph of someone, and the real person in the flesh, and say 'These two faces are the same.' It turns out that our ability to recognise faces is actually the responsibility of a specific part of the brain, which has been named the fusiform face area. It is located approximately above your ears, relatively close to the surface of your brain.

In 2011 Nicholas Furl and colleagues at University College London published a study about a group of people called prosopagnosics. Prosopagnosia is the inability to identify faces, and is sometimes referred to as 'face blindness'. Furl and his team found that the fusiform face area was far less active in prosopagnosics than it was in non-prosopagnosics.[12]

Neurologist Oliver Sacks wrote a very successful book, first published in 1985, which took its title from a case of prosopagnosia – it was called *The Man Who Mistook His Wife for a Hat.*[13] The title comes from a case study that Oliver worked on, where the man had severe impairment in his ability to recognise his own wife. This title initially sounds ridiculous, but one of the key features of prosopagnosia is that individuals who exhibit it have to process faces just like any other type of foreign object, and they generally seem to do so piece by piece. The rest of us have an innate ability to process faces as a whole, but prosopagnosics lack this. We think *this face = Emily*. They may instead think *small nose, big eyes, small ears, familiar voice = Emily*. Apparently about 2.5 per cent of us suffer from a face-processing deficit like this.[14]

In 2009, Richard Russell and his colleagues from Harvard University[15] found that there are also people at the opposite end of the spectrum. According to them, these 'super-recognisers' are 'about as good at face recognition and perception as developmental prosopagnosics are bad'. What is more, this seems to be both a perceptual feature and a memory feature. Super-recognisers sometimes report being able to identify and remember faces years later.

According to one such super-recogniser, referred to as CS, 'It doesn't matter how many years pass, if I've seen your face before I will be able to recall it.' We currently don't know the prevalence of this ability, or exactly how it works. However, because the term 'super-recogniser' has recently gone viral, the research on this ability will likely see some tremendous growth in the next few years.

One area where this ability has obvious applications is policing. People like Josh Davis at the University of Greenwich[16] work with the Metropolitan Police in London to identify and employ such super-recognisers to sift through thousands of photos and make identifications from CCTV footage. In other words, because of their ability they can look at crowds and other complex footage and find a particular face, such as that of a suspect, a task that is incredibly difficult and inefficient for people who do not have this superpower.

There is a test, called the Cambridge Face Memory Test, released by neuroscientists Brad Duchaine and Ken Nakayama in 2004, that can help identify whether or not someone is a super-recogniser. In this test participants see a face from three different angles in the 'study' phase, and then have to identify it out of a line-up of three faces afterwards. This is repeated over many faces. The faces in the test line-ups get more and more similar as the test progresses, making the task more difficult. Super-recognisers are able to correctly identify most of the faces in this test, and in similar tests, and they have been responsible for some tremendous police successes. For example, super-recognisers were heavily involved in identifying people involved in the 2011 London riots, identifying significantly more culprits than facial recognition programs.

This skill is particularly valuable, because apart from such super-recognisers, many people have a difficult time reliably matching faces and photos, as David White from the University of New South Wales and colleagues showed in a 2014[17] study: 'Photo-ID is widely used in security settings, despite research showing that viewers find it very difficult to match unfamiliar faces. Here we . . . ask officers to compare photos to live ID-card bearers, and observe high error rates, including 14 per cent false acceptance of fraudulent photos.' Of course, in line with our lack of insight into our own personal memory biases, most people think that they can identify whether the person standing in front of them is the same person as the one in a photo they are holding. In reality, however, it seems that facial recognition is just another area where many of us are overconfident in our abilities. For most of us, perception and memory seem to interact in a way that can make it difficult for us to solve even the most seemingly basic of identification tasks.

In an ideal world, as well as the police having enhanced abilities to identify suspects, witnesses to crimes would also be able to describe and identify the perpetrator with ease. In such situations the police want us to make assertions with utter certainty. They do not want us to say that the offender *maybe* had a scar, *could have* had brown hair, or was *between* 5'7" and 6'10". The desire for confidence and clarity in identifications of this kind is understandable, but such expectations can also cause confidence judgements to become skewed.

Our intrinsic assessment of the quality of our own memories can come into play here. It may seem an obvious thing to say, but if we think we have a good memory of a person, we generally have high confidence when asked to call on that memory. But as we have seen, just because we *think* we have a good memory of something does not mean we necessarily *do*. So, leaving aside self-assessment, how good are we actually at identifying strangers?

It seems like a straightforward question, but it turns out that there are an almost infinite number of variables that need to qualify this question. How good are we at identifying faces? How good are we at identifying height and body shape? How good are we at identifying people with scars or deformities, people of a different ethnicity, people we only have brief exposure to, older or younger people, people we saw in dim light, people we only paid some attention to, people wearing hats?

In 2013, Matthew Palmer and colleagues at Flinders University[18] conducted a study looking at some of these complexities. They had researchers pair up and take to the streets. Researcher 1 would recruit a participant, and have them provide consent to participate.

Then Researcher 2, who had been hidden, would step into view. Researcher 1 would then tell the participant to look at Researcher 2 until they stepped out of sight once more. Participants were then asked to identify Researcher 2 out of a photo line-up and to rate their confidence in the decision, with half being asked to do so immediately and half being asked to do so about a week later.

As we might expect, participants were better when they made their line-up identification immediately – when they did this they were accurate 60 per cent of the time, and if they made it a week later they were accurate 54 per cent of the time. It may strike you that these rates are pretty low. Indeed, almost half the participants failed to correctly pick the photo of the person they just saw.

What makes this even more disconcerting is the considerable differences in accuracy that result from some pretty basic alterations to a situation. The researchers found that while accuracy and confidence were generally aligned, overconfidence was higher in more difficult conditions. In other words, participants had disproportionately high confidence if they were allowed only a very short viewing of Researcher 2, had a longer delay before their line-up identification, and/or had to divide their attention during the task. So we seem to generally overestimate how good we are going to be at identifying perpetrators in situations where the odds are particularly against us.

Needless to say, this is a complex topic with many amazing people doing amazingly interesting research to help combat both our built-in, and our externally facilitated, memory illusions.

Race face

Another factor that affects our ability to identify others is ethnicity. If you are black and witness an East Asian person committing a crime, good luck getting it right. The same is true for any combination of ethnicities – white, black, East Asian, Indian, Puerto Rican, it doesn't matter. We are simply generally worse at identifying someone of a different ethnicity, a phenomenon known as ORB – own-race bias. These cross-race effects present a tremendous problem for the legal system, as clearly people of different ethnicities commit crimes against one another. And what is worse, the system is constantly battling what looks like everyday systematic racism.

Perhaps we *are* all just racist, even if we don't admit it. Or, perhaps there is something else going on here. ORB has been heavily studied, and one of the prevailing assumptions is that it has to do with how we remember faces.

According to Caroline Blais and colleagues from the University of Glasgow,[19] culture shapes how we look at faces. In 2008 they published research that demonstrated this using eye-tracking technology to give an indication of what a subject was looking at, and therefore presumably processing. The participants in the study were presented with photos of Western Caucasian and East Asian faces to look at. All the participants self-identified as Western Caucasian or East Asian themselves.

They found that if the participant was Caucasian, they generally used a triangular pattern to look at the faces. They looked at eyes, mouth, nose, and then explored other parts. For East Asian participants there was a different pattern; they focused far more centrally, seemingly mostly looking at the nose. And they did this regardless of the ethnicity of the person in the photo. The researchers interpret this as a culturally derived, suggesting that 'direct or excessive eye contact may be considered rude in East Asian cultures and this social norm might have determined gaze avoidance in East Asian observers'.

Alternatively, different search strategies may focus on things that are typically varied within a certain culture. For example, it makes sense to spend time encoding eye colour in Western Caucasian groups, since there is significant variation, while it may not be as useful as a distinguishing feature for non-Caucasian groups. Whatever the underlying reason, this research demonstrated that participants had culturally influenced facial search strategies that could lead them astray when looking at or identifying foreign faces.

So, inappropriate allocation of attention to particular facial features appears to be one of the main reasons for the existence of ORB, at least according to Blais and her team. Focusing on the 'wrong' characteristics makes it harder for us to identify and remember exactly what a person looked like, and we are more likely to do this if the person is a different race from us. This flows into other person characteristics as well. For example, in ethnicities that have less variation in hair colour or height, encoding these is hardly useful.

All this ties in quite naturally and cyclically with memory. According to an article published in 2014 by David Ross and colleagues at Vanderbilt University,[20] the reason we are able to recognise faces at all is because we have these face-learning strategies. Ross suggests that it is because we have a strong set of memories of what faces look like that we are able to identify new faces. More specifically, he suggests that faces are represented in our brains by their similarity to exemplars of previously experienced faces. In other words, we remember new faces in relation to our database of faces that we already have: *How similar is this new face to old faces?*

This so-called exemplar-based model of remembering faces means that our existing memory bank matters. It allows us to optimise the way we analyse faces, minimising the time and effort we must invest in learning a new face. However, these strategies that we apply then taint our memories of new faces, and can backfire when a face has too many new features. It's almost as if our face database can't handle too much newness at once. Luckily for us, while we may initially be bad at recognising faces from other ethnicities, each new face we encode results in a small update to our database of what faces look like. This relates to an idea known as the contact hypothesis, generally attributed to American psychologist Gordon Allport in 1954. This is the idea that people generally become more understanding of other groups, and more appreciative of their points of view, the more contact they have with them. This may also apply to facial recognition. Indeed, according to a review by Steven Young at Tufts University and his colleagues in 2012,[21] there is some, albeit mixed, evidence to support the idea that the more exposure we have to other-race faces, the better we get at identifying them.

In addition to the own-race bias, we have also been shown to have an own-age bias and an own-gender bias. Siggi Sporer[22] from the University of Giessen suggests that this is because *otherness,* in general, is bad news for memory. In a review conducted in 2001, Sporer argues that not only are we bad at identifying people from other groups of any kind, but we are generally overconfident in our ability to perform well at this. As with many other tasks discussed in this book, we think we are going to be good at identifying others, even if we think others may not be good at identifying us.

So, while eyewitness identification is at the core of most legal trials, research shows that there are fundamental memory characteristics that make any such identification a whirlwind of possible errors. In such cases, independent pieces of corroborating evidence are needed if we are to feel at all confident that an identification is correct. The Dutch have a great saying that applies here: one witness is no witness.

Making monsters

When I am not lecturing or conducting memory research, I sometimes work on criminal trials. These typically involve issues related to the memory and identification topics we have discussed up to now. Since lawyers and the police generally only bring in a false memory expert when something very bad has happened, I tend to work on the kinds of highly unpleasant cases that make people worry about the fundamental nature of humankind. Murder. Abuse. Sexual assault.

While most cases I work on are shocking, nothing has ever worried me as much as my first. Note that I am going to obscure the details for the sake of the integrity of the case. In this case it was alleged that a number of teachers and members of the clergy had abused – both sexually and non-sexually – a large number of the pupils at a religious-run school. It was a historical case, so these crimes were said to have taken place over 40 years previously. The case had already been investigated twice before, both times being abandoned due to insufficient evidence for criminal proceedings, leaving a shelf filled with binders of documentation to be waded through. Some of the alleged perpetrators and witnesses had already passed away. The police staff had changed multiple times since the case was first opened, and none of the detectives who had originally worked on the case were still on it. It seemed as though everyone had to start from scratch. Well, everyone except for the people leading the charge on behalf of the victims, as they had told their side many times to many people over the years, as police came and went.

I had been contacted by a detective at the police force that was dealing with the case, who had asked me to stop by to talk about ways I might potentially be able to become involved with their work. I was both pleased and impressed because that sort of thing does not happen very often – we academics can seem inaccessible to the very professionals we would like to help. However, here was the detective asking about my research.

'We need someone to help us catch these monsters,' he said.

'These *suspects*', I corrected him.

In a world where catching the bad guys is the motivation to get up every day and deal with tremendously difficult people and situations, using a term like 'monsters' makes sense. Of course they want to catch *monsters.* But in some situations we do not know who the bad guys are, or whether they exist at all.

This was one of those situations. Of those involved in the case, the overwhelming majority claimed that nothing untoward had ever happened to them, at least nothing outside the bounds of acceptable forms of punishment that were available to teachers at the time – those were the days when it was legal to use basic forms of physical punishment. Only a very small group of former pupils claimed to remember bits and pieces of unusual, problematic situations. What made things even trickier was that these bits and pieces seemed to have drastically changed and grown over time.

Still, I was armed with my notepad and a pen, determined to establish *who said what to whom when.* Just because a case is tricky, and memories and accounts look contradictory at first glance, this does not mean that that the case should be dismissed – it just means that people need to be aware that allegations are being dealt with, not facts. Only if there are too many red flags, and insufficient independent corroborating evidence, might we begin to think there is reasonable doubt and that a case is based on false allegations or false memories.

This particular case is still ongoing, and it will likely take years to figure out whether the claims are reliable. So far, there are a number of red flags, including victims' accounts drastically changing over time, and denial by most of those who were at the school that any of this happened at all. A large number of denials, combined with not much corroborating evidence, will make it a difficult case. Either way, it is important for police in such situations to at least acknowledge that all of us can succumb to tunnel vision and not to get too caught up in the assumption that there *must* be monsters to catch.

Making Monsters is an excellent book by sociologist Richard Ofshe and Ethan Watters[23] which highlights the way confidence and assumptions about memory can lead to the personal acceptance of false memories of victimisation. Processes involving cascades of such assumptions can take hold in the justice system, radically affecting the course of a case; one overconfident witness or victim can start a domino effect which, in the worst-case scenario, could end up putting innocent people in prison.

It is crucial that the justice system becomes aware of overconfidence factors, memory illusions and the problems we have with identification, since these can lead to atrocious situations where we sometimes build cases based on nothing but air. There is a tendency when a crime has been committed, especially a particularly horrific one, to assume guilt of an accused party. While of course it would be terrible for someone guilty of a terrible crime to go unpunished, it is surely equally terrible to end up punishing the innocent as a result of poor practice or inadequate understanding.

It is tempting to think that perhaps the police, under pressure to secure convictions, might be more vulnerable to making such errors than the rest of us. That's simply not the case – we are all defective detectives when it comes to identifying real memories or true accounts of events, be they of crimes or everyday occurrences. We are all vulnerable to the same kinds of memory and confidence illusions. And we need to realise that confidence is *not* key in these kinds of situations. To me, high confidence is often instead a warning sign. WARNING, *this person may not fully appreciate their biases.* WARNING, *this person may not be aware of memory illusions and shortcomings.* WARNING *this memory is too good to be true.* I approach high levels of confidence with high levels of caution because if it is *over*confidence, it can be incredibly destructive.

7. WHERE WERE YOU WHEN 9/11 HAPPENED?

Flashbulbs, memory hacking, and traumatic events

Why our memory for emotional events is flawed

In 2015, Brian Williams, the most-watched television news anchor in America, was suspended from his job at NBC Nightly News amid confusion and recriminations. In 2003, Williams had gone to the front lines of the war in Iraq to give a news report and while he was under way his helicopter platoon came under fire. Ten years later he described the event during a TV interview with David Letterman:

> Two of our four helicopters were hit by ground fire, including the one I was in. RPG and AK47 . . . We were only at 100 feet doing 100-forward knots. We landed very quickly and hard and . . . we were stuck, four birds in the middle of the desert and we were north out ahead of the other Americans. . . . They started distributing weapons and we heard a noise. It was Bradley fighting vehicles and Abrams tanks coming. They happened to spot us. This was the invasion. The US invasion. They surrounded us for three days during the sandstorm that was so big that it suspended the war effort. It was called 'Orange Crush'. And they got us out of there alive.

Two years later, in 2015, he reflected upon the situation once again whilst on air:

> The helicopter we were travelling in was forced down after being hit by an RPG. Our travelling NBC News team was rescued, surrounded, and kept alive by an armour mechanised platoon from the US Army 3rd Infantry.

This is a very intense story with many specifics. Williams repeated it on numerous occasions and clearly had enough confidence in it to relate it on television for everyone to see. And everyone did see it, including the men who were present in the helicopter that went down. 'Sorry dude, I don't remember you being on my aircraft,' wrote one of them on the NBC News Facebook page,[1] in response to a video of Brian Williams describing the event. Another responded to the Facebook post: 'He was actually on my aircraft and we came in behind you about 30–40 minutes later.'

It transpired that Brian Williams had told the story of what had happened to an aircraft in front of him – he was never in the helicopter that was attacked, and since it was an event with so many witnesses his account was easy to debunk. A media storm ensued. Everyone immediately assumed that Williams had purposely embellished his experience in Iraq in order to bolster his own reputation. He apologised, but the damage was done and his credibility was shattered.

However, given the nature of the work that I do, I cannot help but try to contextualise the immediate jump to the conclusion that he was making it up. To me it seems premature to pull out the pitchforks when we aren't sure why someone is giving an inaccurate account, mostly because we can unfortunately never actually tell unintentional and intentional fabrication apart, unless the person later tells us that they were lying. What it definitely does bring to light, however, is a core assumption that many people make about memory. To me it very much seems like what happened in the Chopper Whopper scandal was that Williams was accused of being a liar, at least in part, because our general assumption is that no one could possibly misremember such an emotional event. But is that really the case?

Highly emotional

If you are like most people, you probably believe that trauma memories are *special*. But your understanding of these trauma memories is probably self-contradictory. You probably believe that on the one hand we often forget or repress incredibly emotional events, but on the other that we can have nightmares and flashbacks to them. If this is true, then you probably believe that our memories of traumatic events are simultaneously worse and better than non-emotional memories. But, which is it?

In a 2001 paper aptly entitled 'Is traumatic memory special?', Stephen Porter at Dalhousie University and Angela Birt at the University of British Columbia suggest[2] that there are a number of different viewpoints when it comes to memories of highly emotional events.

The first of these, they say, is the traumatic memory argument, which is the view that we remember traumatic events differently, and often worse, than other kinds of events. The underlying assumption behind this argument is that the high emotional impact of traumatic situations overrides our other processing abilities. Proponents of this viewpoint would argue, for example, that a soldier in a war zone may be so severely shell-shocked that they have trouble encoding or recalling a coherent memory of a battle. We can actually trace this idea all the way back to Aristotle:

'Memory does not occur in those who are in a rapid state of transition, whether owing to some perturbing experience or their period of life; just as no impression would be left if a seal were stamped on running water.'

This view assumes that our memories of traumatic events are thus stored as fragmented images, emotions and sensations without a coherent structure. A soldier might remember the smell of the battlefield, the sound of gunshots and the taste of blood, but not remember any particular events. Proponents of the traumatic memory argument claim this is why PTSD sufferers sometimes experience powerful flashbacks – they are remembering little fragments of traumatic memories rather than whole events.

On top of the idea that we may store such memories as individual fragments without a structure, proponents of the traumatic memory argument also often contend that people can 'dissociate' during highly emotional events. The word dissociation is used in many ways, but it most often describes symptoms including derealisation, feeling as if the world is not real, and depersonalisation, feeling as if you are not real.[3] Such dissociation can supposedly occur during a highly emotional event, when a person feels like they are not actually there, or after such an event, when a person has persistent feelings of not being real.

Proponents of the traumatic memory argument, such as professor of applied psychology Judith Alpert, who in 1998 formed a working group with colleagues to investigate the nature of trauma memories,[4] suggest that dissociation is 'a psychological defense against the impact of trauma and the mental mechanism that most likely accounts for the amnesia and hypermnesia commonly experienced by traumatised individuals'. In other words, things can be so traumatic that our mind has trouble dealing with them at all, essentially making such memories inaccessible. The statement is also inherently contradictory because it simultaneously says that we have both amnesia, which is severe forgetting, and hypermnesia, which is enhanced recall of an event.

From my own experience, the belief that this can happen is widespread amongst both practitioners and the general public. But, is the assumption of dissociation as a response to trauma reasonable? This is an important question, as according to medical doctor Angelica Staniloiu and her colleague Hans Markowitsch in a review of the science of dissociation in 2014,[5] 'dissociative amnesia is one of the most enigmatic and controversial psychiatric disorders'.

Researchers such as Porter and Birt say that claims in support of the traumatic memory argument are poorly supported. Most memory researchers since the early 2000s argue that while dissociation may be possible, people usually do not dissociate during emotional events, and that there is no evidence to support a special fracturing process of memory in trauma situations. It is also unlikely that there is such a thing as repression, hiding emotional memories from direct access, but we will get back to that in a later chapter.

Related to, but different from, the traumatic memory argument is the medical argument of trauma encoding. This also suggests a diminished memory for such events but is different because it focuses exclusively on principles of brain biology. It includes the basic premise that accidents or violent assaults may cause physical trauma to the brain, which can result in amnesia. In other words, actually physically damaging parts of the brain can lead to memory loss.

The medical argument also includes the idea of *mnestic block*. This posits the existence of a type of amnesia due to problems in the way the brain works rather than physical damage. According to Hans Markowitsch from the University of Bielefeld and his colleagues[6], mnestic block syndrome is 'related to an altered brain metabolism which may include changes in various transmitter and hormonal systems (GABA-agonists, glucocorticoids, acetylcholine)'.

These sorts of medical impairments, unlike the vague traumatic memory argument, are not thought to cause fragmented memories, and rather than leaving a memory gap for the single inciting event, are generally thought to cause weeks or even years of amnesia. They are also scientifically sound and medically indisputable; breaking the parts of your brain responsible for memory processing, whether it be in structure or function, *will* disrupt your memory. However, such specific disruptions to the brain's memory system are very rare, barring pathological memory-impairing conditions that befall many of us in later life, such as Alzheimer's or dementia.

In spite of assertions about the negative impact of trauma on memory, the majority of current research actually supports the idea that, barring a broken brain, there is a 'trauma superiority effect' on memory. In 2007 Stephen Porter and Kristine Peace at Dalhousie University[7] published a study that looked into this issue. They recruited participants who had recently experienced trauma, and interviewed them immediately, then 3 months later, and then about 3.5 years later. They asked them about their traumatic memories and also about highly positive emotional memories at the same time, focusing on features such as vividness and clarity, the overall quality of memory relative to other memories, and the existence of sensory components to the memory such as sights, sounds and smells.

Porter and Peace found that the memories of the traumatic events were highly consistent over time, remaining virtually unchanged in almost all features. It also turned out that in comparison to the highly positive life experiences, the memories of negative experiences were significantly more stable over time.

These findings suggest, as does Porter and Birt's research, that memories of trauma *are* special, just not special in the way that many people assume, indeed often being better than other kinds of memories.

This is supported by a 2012 review of the literature by the Svein Magnussen and Annika Melinder from the University of Oslo in 2012:[8] 'The current evidence from systematic and methodologically sound studies strongly suggests that memories of traumatic events are more resistant to forgetting than memories of mundane events.' This is both good news and bad news, as it seems to increase the likelihood of accuracy for eyewitness and victim testimonies (though they will still be up against all the problems discussed previously), but it also means that traumatic memories we might rather forget could haunt us forever. Interestingly, this persistence of memory seems to occur not only with traumatic events we directly experienced ourselves, but also indirectly experienced traumatic events that we have heard about repeatedly through the media.

Flashbulbs

'Where were you when 9/11 happened?' was a tremendously popular question in the 2000s. Similarly, in the decades before, people might ask each other's whereabouts when the Challenger shuttle exploded or when JFK was shot.

What these types of questions imply is that we have the capacity for immediate powerful recollections of the circumstances we were in at particular significant moments. These are sometimes referred to as *flashbulb memories*. They are detailed and vivid, and typically involve recalling the situation in which a piece of historically important news was heard, along with a detailed recollection of the event itself. Rememberers often mention the informant who shared the news, what they were doing when they heard the news, what they were wearing, and what they were thinking, feeling and saying. I mentioned this phenomenon in a university lecture recently, and one of the students in the audience said that on finding out that the first plane had flown into the twin towers in New York on 11 September, 2001, she had made a crude joke about the inability of pilots to fly straight – a joke she had felt self-conscious about ever since.

In 1977 Harvard University researchers Roger Brown and James Kulik[9] investigated these kinds of memories. They sent out a questionnaire to 80 people to ask about what made them remember important historical events such as assassinations, highly newsworthy occurrences and personally important experiences. From the questionnaire responses, they concluded that many people have memories with considerable perceptual clarity for important historic events. In other words, people could report more correct details with higher confidence for certain kinds of events, with these events having three main characteristics.

First, the event needed to generate a high level of surprise. It could not be a trivial or expected event. Second, the event needed to carry important consequences for the person or for people in general – referred to as having a high level of consequentiality. This can be an important consequence for them personally, or for society. For example, 9/11 may not be personally relevant to a particular individual, but it was quickly deemed important for society, making it an event with a high level of consequentiality. Finally, the event had to generate high levels of emotional arousal – the individual needed to experience fear, sadness, anger or some other strong emotion. Brown and Kulic argued that without these three conditions being fulfilled a flashbulb memory could not occur.

They went on to speculate (without, it must be said, any reasonable scientific backing) that the reason we can have these kinds of memories is because of a unique biological mechanism that creates a permanent hard-wired record of the event in the brain. According to them, rememberers of flashbulb memories display extremely high confidence, often declaring the details with definitive statements – 'I was *definitely* at home.' 'I remember it *so clearly*.' Take these examples from the False Memory Archive, a project by the artist and Wellcome Trust Engagement Fellow A. R. Hopwood:[10]

> Sam: 'I have a very vivid memory of watching the Challenger shuttle disaster in high school. I was standing in the Science/Media library in my school, and there was a TV on a tall metal rolling stand – the kind that was rolled into classrooms for presentations. Several of my friends were there. I can remember exactly how I was standing in the room, where I was, the angle I was seeing the screen from, the shock of others in the room.'

> Sue: 'I remember my housemate coming home to our shared house, telling me that she heard on the radio that a plane had crashed into the World Trade Centre. I have a *clear* memory of that house and that she went to her bedroom, and I turned on the TV news.'

Both those recollections are extremely vivid and detailed, and are told with a confidence and conviction that gives them the ring of truth. In 2014 Martin Day and Michael Ross at the University of Waterloo[11] published a study that further explored rememberer confidence in flashbulb memories. Instead of a questionnaire, they conducted interviews. The first interviews took place shortly after the death of Michael Jackson, and they asked participants to recall how they found out about the death, where they were, their confidence in their memories, and whether they thought their memories would be durable. Eighteen months later, they asked the participants to once again recall the details of their whereabouts when they found out about the death, and asked them once again about their confidence in the memory.

And what did they find? I'm afraid I have to pull the rug out from under you. The researchers found that these memories often lacked consistency, the descriptions of them changing between the two memory interviews, despite confidence in the accuracy of the memory remaining very high. This suggests that flashbulb memories may not be as permanent or accurate as Brown and Kulic originally proposed, and that people are overconfident in the accuracy of their recall for these kinds of events – which supports the idea covered in earlier chapters, that confidence does not necessarily imply accuracy. The original Brown and Kulic study had significant methodological problems, including having a small sample, relying only on self-report, assuming that participants' reported memories were accurate, and making unfounded assumptions about how the brain works. Despite pitfalls, their research led to a large amount of further study of these flashbulb memories, resulting in an erroneous sentiment that memories for important historical events are safe from corruption. In reality, memories of witnessing cultural events may not be as strong and protected as we often assume. To demonstrate that, let's revisit those accounts from Sam and Sue, which I hope you will agree were detailed and plausible. The problem is, I'm afraid they both had a little more to say:

Sam: 'The problem? The Challenger disaster happened two years after I graduated. When it occurred, I wasn't in that high school, or that city – I was living in another part of the country. This memory is completely real to me, and yet I know that it's false – it didn't happen that way. I have no idea where I really was when, or if, I saw the Challenger disaster on TV – but I know for sure I wasn't where I "remember" being.'

Sue: '. . . we had stopped living in that particular house three years before 9/11. At the time of 9/11 we were sharing a house again, but in a different house. My memory has mixed up the death of Princess Diana and 9/11; same housemate, but wrong house.'

This process of realising that a particular memory is inaccurate or impossible is called recollection rejection, a term coined by memory scientists Charles Brainerd at Cornell University and his colleagues in 2003.[12] What is amazing about instances when we reject our memories in this way is that it does not necessarily mean that we will no longer have the memory, it simply means that our confidence that it happened diminishes greatly – or disappears altogether. More often than not, however, we are not confronted with contradictory evidence for our memories, and can come to accept accounts as our personal reality, even if they make no logical sense.

Regarding trauma memories, there is also potential for the integration of grossly false information about events that we have, or think we have, experienced ourselves. Even our highly emotional memories can be totally false. How do I know this? Because it's part of my job as a researcher to show that even our most vivid memories may be up to no good.

Memory hacking

I am a memory hacker. I get people to believe things that never happened.

Usually when I tell people that this is part of my job they ask 'But why?' The answer is that I believe that through generating memories in a lab-based experiment we may be able to begin to figure out how such memory illusions operate. We cannot hope to prevent such errors from happening until we fully understand why they occur.

I'll explain what I actually do. It's not hypnosis, or torture, or anything like that – it's simple social psychology. I apply what has been learned from decades of research, and basically do the reverse of the careful, non-leading questioning which I supervise when working with the police. I deliberately provide what I think are the ideal conditions for the development of memory illusions. Let me take you through the process, one step at a time.

Step 1: I source adult participants for an 'emotional memory study', and get them to provide contact information for some informants – people who know them really well, such as their parents.

Step 2: I contact the informants and ask them to describe emotional experiences that the potential participant may be able to recall from a particular time in their lives, between the ages of 11 and 14. At this point I also harvest information about who the potential participant was best friends with at the time and where they lived.

Step 3: I filter the participants so as to include only people who have not actually experienced any of the emotional events I am planning to implant, but who have experienced at least one other emotional event. I invite these people to participate in the study.

Step 4: The participants come in for the study. They are under the impression that it is a study on emotional memory, but have no idea that the actual premise includes the possible implanting of false memories. I get their consent to participate in an emotional memory study. The deception I use, in that I don't disclose that one of the memories they may recall will be false, is of course in compliance with the permissions granted to me by my university research ethics board. I begin to ask the participant, in a structured way, about their memory of a true emotional event which I have learned about from the informants. It might be being bullied at school, fainting on vacation or any other emotional event. As a result of this I gain credibility as someone who knows about emotional memories that they have experienced.

Step 5: I introduce a false event, telling the participants they did something that I know they did not actually do. The event in my most recent experiment, published in 2015 with Professor Stephen Porter and conducted at the University of British Columbia,[13] involved telling participants that they had committed a crime with police contact – assault, assault with a weapon, or theft – or had experienced another emotional event – an animal attack, a bodily injury, or losing a large sum of money and getting into trouble with their parents. The script we used for the police contact scenario was as follows:

> Okay [participant's name], thank you for telling me about the first event. You did a great job. The other event your parent[s] reported happening was an

> incident where you were in contact with the police. I will be asking you about this next. On the questionnaire, your parents stated that when you were about [age] years old, you [false event]. It happened in [place], in the fall and you were with [friend or relative] when it happened.
>
> No other information was divulged about the supposed event.

Step 6: After being told that this supposedly happened in the first interview, participants initially correctly say something along the lines of 'I don't remember this', so I offer to help them out. I offer to do a visualisation exercise with them. In this exercise I get the participants to close their eyes and picture what the event would have been like. Little do they know that I am doing this to get them to access their imagination instead of their memory. After they do this visualisation exercise during the first interview, they normally don't have many details. At this point I send them home with the instruction that they are not allowed to talk to anyone about the study, that they should try to visualise the memory at home, and that they should come back to see me in a week's time.

Step 7: The participants return to my lab a week later, and I ask them again to take me through their recollections of the true memory. Then I ask them about the false event. At this point, many participants begin to 'remember' and report details of it. 'Leaves were falling. Blue skies. I stole a CD. I punched a girl for teasing me. The policeman had brown hair.' I encourage them and say they are right on track – positive reinforcement. I also repeat the visualisation exercise, to get them to imagine more details that they can mistake for memory details. I send them home again, tell them to try to get more details, and to come back for a third time, a week later, where I repeat the process – true memory recollection followed by false memory recollection with visualisation – for a final time.

Step 8: After three interviews, I harvest the full-blown false memories. At this point many of my participants are divulging a tremendous number of details about an event that never happened, talking about them with confidence. It's like memory magic.

You may not think this would work on you, but the statistics suggest otherwise. In this particular study we found that 70 per cent or more of participants in both the criminal and other emotional conditions developed full false memories. I defined full false memories by a number of criteria, including the participant at least reporting ten details related to the event, and saying at the debriefing that they believed that the event actually happened. Many of the accounts given were richly detailed. Here an excerpt from one participant's transcript; we'll call her A:

> I remember being so shocked when the cops came. That was bad. That was bad.

What was so bad? Here is an earlier part of her account:

> That's what set me off, [she] was calling me a slut . . . Because I was a virgin . . . I know she wasn't close to me. And I think we were like, we were like, following her, and taunting her . . . And I got pissed off. And it wasn't a huge rock. It wasn't . . . but I did find it, it was a decent sized rock and I threw it at her head. I threw

> it at her . . . I ended up going home. S took off. And then, I remember being in the house, I think we were eating dinner, and uh, and then we . . . the doorbell rang and my mom went and answered the door, and then I remember mom yelling for me to come to the door, so I went, and there's two cops standing there.

You might still not be convinced. You may think only vulnerable people would fall for this. Certainly, if we use poor or deliberately misleading interview techniques on individuals who are highly compliant, very young, sick, or low functioning, we might expect these kinds of results. In this study, however, I specifically selected participants who were not vulnerable in any of those ways. They were ordinary university students.

That means that the most compelling explanation for the results we got is that even people like these are susceptible to social pressures and bad memory retrieval techniques that helped them to imagine things, and helped them to mistake those imagined things for real experiences. This is all in line with ample research conducted by other teams, who have successfully implanted other kinds of emotional false memories such as spilling a bowl of punch on the bride's parents at a wedding, published by Ira Hyman and his colleagues at Western Washington University in 1995,[14] or of being attacked by a vicious animal, as generated by Stephen Porter and his colleagues at the University of British Columbia in 1999.[15]

In two follow-up studies I conducted[16] I showed videos from my false-memory-of-crime study to new participants. These new participants did not know that some of the videos they were asked to watch involved false memories. I had each participant view a recording of the same person recalling first a true memory and then a false one. In both studies, participants were no better at correctly guessing the true and false memory than if they had picked randomly. The evidence seems to be that these memories feel real to rememberers, and they therefore look real to others – they can become part of the rememberer's personal past, whether they actually happened or not.

In the wild

But perhaps you're still not convinced. Perhaps you think fake memories generated in a lab situation don't prove anything about our experiences in the real world. If that's your opinion then you are not alone – which is why false memories have also been studied 'in the wild'. Rather than creating scenarios themselves, researchers can piggyback on real-life events that they know already cause the kinds of conditions they require – being highly negative and emotional, generating a level of stress that we should never seek to generate in a lab environment.

An example of this kind of highly stressful situation is navy training. Let's imagine the scenario. You are a 26-year-old member of the US Navy, and as part of your survival training you've been put in a simulated prisoner-of-war situation. You have just finished a four-day evasion exercise in the wilderness. You are tired, you are sore, you are hungry. And now, much to your dismay, you find yourself held captive in a prisoner-of-war camp. Technically you know this is a mock scenario, but the types of stresses you are going to be subjected to are directly modelled on the experiences of actual prisoners of war.

For the interrogation you are locked into a room with someone you have never met, one-on-one. You undergo a vicious half-hour where you are subjected to physical abuses intended to make you talk, including slaps to the face, abdominal punches, being slammed into the wall, and being placed in stress positions. You are required to stare directly into your interrogator's eyes for much of the ordeal. His face is not covered. You can clearly see him at almost all times. Then he places you in isolation. During this prisoner-of-war exercise you are placed under acute stress for a total of 72 hours. This is, by most people's standards, a potentially traumatic experience.

Given that identifying the interrogator could be essential intelligence that the navy would want to know should you get released, and since you have specifically been trained to bring home this kind of information, you paid a lot of attention to his features. You certainly had enough time to look at him. So, if I put two photographs in front of you, do you think you would be able to identify which showed your interrogator?

In 2013, PTSD researcher Charles Morgan from Yale University and a number of colleagues published a study[17] that examined whether individuals in this exact situation could be prone to the same kinds of misinformation effects that have been observed in lab studies.

How did they do it? Simply by showing the mock prisoners a single mugshot photograph for a few minutes while they were in solitary confinement. The photo was not the interrogator, but was introduced by the person who brought it in as if it were. The original interrogator had medium-length curly brown hair and round features, while the person the participants were shown had no hair and very narrow features. They looked totally different. In spite of this, when the participants were subsequently asked to identify their interrogator, the overwhelming majority – 84 to 91 per cent – misidentified them and picked the false photo. The researchers had intentionally introduced misinformation, and it had replaced the participants' memory of the real face.

In the same study Morgan and his colleagues also demonstrated that they could change whether participants reported the presence of neutral items such as glasses or a telephone in the room, or even more critical pieces of information such as the presence of a uniform or weapon. Depending simply on how the participants were asked about their experiences when they returned to the barracks, the presence or absence of these features in their memories underwent marked changes. When leading questions were used, such as 'Was the uniform worn by your interrogator green with red boards or blue with orange boards?' or 'Did your interrogator allow you to make a phone call? Describe the telephone in the interrogation room', 85 per cent and 98 per cent of participants respectively described that they saw the uniform, and that there was a phone in the room. It's true that even when no misinformation was presented some participants included inaccurate details, but this was comparatively rare. In general it seems that simply being shown photographs or asked particular questions can plant false details in our memories, even for incredibly emotional events.

Better left unsaid

It's not only outside sources that can dramatically alter our recollections of emotional events; we are also prone to distortion from internal influences. One way this can happen is through sharing our memories with others, something that most of us are likely do after important life events – whether it's calling our family to impart some exciting news, reporting back to our boss about a big problem at work, or even giving a statement to police. In these kinds of situations we are transferring information that was originally encoded visually (or indeed through other senses) into verbal information. We are turning sensory inputs into words. But this process is not flawless; every time we take images, sounds or smells and verbalise them we potentially alter or lose information. There is a limit to the amount of detail we are able to communicate through language, so we have to cut corners. We simplify. This is a process known as verbal overshadowing, a term coined by psychological scientist Jonathan Schooler.

Schooler, a researcher at the University of Pittsburgh, published the first set of studies on verbal overshadowing in 1990 with his colleague Tonya Engstler-Schooler.[18] Their main study involved participants watching a video of a bank robbery for 30 seconds. After then doing an unrelated task for 20 minutes, half of the participants spent 5 minutes writing down a description of the bank robber's face, while the other half undertook a task naming countries and their capitals. After this, all the participants were presented with a line-up of eight faces that were, as the researchers put it, 'verbally similar', meaning that the faces matched the same kind of description – such as 'blonde hair, green eyes, medium nose, small ears, narrow lips'. This is different from matching photos purely on visual similarity, which may focus on things that are harder to put into words, such as mathematical distances between facial features.

We would expect that the more often we verbally rehearse and reinforce the appearance of a face, the better we should retain the image of it in our memory. However, it seems that the opposite is true. The researchers found that those who wrote down the description of the perpetrator's face actually performed significantly worse at identifying the correct person out of the line-up than those who did not. In one experiment, for example, of those participants who had written down a description of the criminal, only 27 per cent picked the correct person out of the line-up, while 61 per cent of those who had not written a description managed to do so. That's a huge difference. By rehearsing only details that could be readily put into words, the participants had de-emphasised the nuances of their original visual memory, making it harder to access.

This effect is incredibly robust, as indicated by quite possibly the biggest replication effort of all time in psychology.[19] This was a massive collaborative effort by almost 100 scholars and 33 labs, including Jonathan Schooler and Daniel Simons, and published in 2014. All researchers followed the same protocol, and they found that even when the experiment was conducted by different researchers, in different countries, and with different participants, the verbal overshadowing effect was constant. Putting pictures into words always makes our memories for those pictures worse.

Further research by Schooler and others has suggested that this effect may also transfer to other situations and senses. It seems that whenever something is difficult to put into words, verbalisation of it generally diminishes performance. Try to describe a colour, taste or music, and you make your memory of it worse. Try describing a map, a decision or an emotional judgement, and it becomes harder to remember all the details of the original situation. This is also true when others verbalise things for us. If we hear someone else's description of a person's face, a colour or a map, our memory of it is also impaired. Our friends may be trying to help us when they give their verbal account of something that happened, but they may instead be overshadowing our own original recollections.

According to Schooler,[20] besides losing nuances, verbalising non-verbal things makes us generate competing memories. We put ourselves into a situation where we have both a memory of the time we *described* the event and a memory of the time we actually *experienced* the event. This memory of the verbalisation seems to take precedence over our original memory fragment, and we may subsequently remember the verbalisation as the best account of what happened. When faced with an identification task where we need all the original nuances back, such as a photo line-up, it then becomes difficult to think past our verbal description. In short, it appears our memories can be adversely manipulated by our own misguided attempts to improve them.

This does not mean that verbalising is *always* a bad idea. Schooler's research also shows[21] that verbalising our memories does not impair performance – and may even enhance it – for information that was originally in word form: word lists, spoken statements, or facts, for example.

Another way we try to hang on to the past is by taking pictures. We tell ourselves we are making memories, that these photos will help us remember our lives. But if we can overshadow our memories by verbalising them, can we also overshadow them with pictures? Of course, photos inherently retain some of those nuances that we lose when we verbalise things, but the issue of potentially creating competing memories remains. In 2011, Linda Henkel at Fairfield University conducted research[22] to examine the impact viewing photographs can have on memory. She had participants complete a series of tasks such as breaking a pencil, crushing a paper cup, or opening an envelope. She then had them return a week later and simply asked them to match photos with descriptions of tasks. Some of the photos were of the tasks the participant had completed, while others were not. Two weeks later the participants returned again, and this time they were asked to identify which of 80 tasks described on a list they had completed during the first part of the experiment.

Simply seeing photos of completed tasks made it more likely for participants to think that they actually had completed them, even without the researcher making that implication – they thought they had done things simply because they saw photos of them. Seeing a photo made participants about four times more likely to say they had done something they had not done.

This effect extends to more complex autobiographical experiences as well. Research from 2008 by Alan Brown at Southern Methodist University and Elizabeth Marsh at Duke University[23] has demonstrated that simply showing people photos of particular locations makes them more likely to erroneously report having visited those places when asked a week or two later. Participants were more likely to misremember visiting places that were mundane than unique places. Because their study was investigating memory for visiting locations on a college campus, mundane locations included things that exist on all campuses, including classrooms, libraries and streets. Unique locations included photos of statues, artwork and particularly ornamental buildings. When questioned, 87 per cent of their participants claimed to have visited at least one mundane location and 62 per cent claimed to have visited at least one unique location. None of the photos were from the campus the students actually visited, they were from an entirely different campus, so it was impossible for the students to have seen the depicted locations during the campus tour. A possible explanation is that it is easier for us to picture and accept visiting places that are mundane, since we can draw on real previous memories of similar places that we can mistake as a memory of the fake visit.

When researchers manipulate images or introduce misinformation to suggest that people did things that they never did, the problem unsurprisingly becomes even worse. In 2002, research by Kimberly Wade and Maryanne Garry at the Victoria University of Wellington, along with their colleagues Don Read and Stephen Lindsay from the University of Victoria, showed[24] that half of the participants in a study could come to recall details of a hot-air balloon ride they had never taken simply through being asked to remember the supposed event while being shown a photoshopped image of themselves in the balloon basket.

Another study, published in 2004 by Stephen Lindsay from the University of Victoria and his colleagues,[25] showed that the photos didn't necessarily even have to be altered. The team had half of their participants imagine experiencing three events from childhood, while the other half were asked to do the same while looking at a real photo of their former school classmates. Participants were then asked to recall their memories of the events in question. Two of these had actually happened (information about these true events had been provided ahead of time by the participants' parents) but the third was a fictional event which had been invented by the team.

Of those who were asked to picture the event happening, 45 per cent formed false memories of it, while an astonishing 78.2 per cent of those who pictured the event *and* were exposed to true pictures of old classmates formed false memories. So giving pictures to the participants who were trying to remember events made them more likely to create memories of things that never actually happened. These real pictures served as a foundation that the participants could meld into their false accounts, making them feel more real.

It seems that photos can quite severely mislead our memories, especially when coupled with deliberate misinformation. One of the main reasons for this is presumably similar to the cause of verbal overshadowing; when we see a photo we create a new memory of that occasion which can interfere with our memory of actually experiencing (or not experiencing) an event. When we think about the event we may then have trouble distinguishing between our memory of the photo and our actual experience – possibly even entirely replacing a real visual memory with another. Emotional or not, verbal or visual, our memories can be readily manipulated.

Critical incident stress debriefing

Given all this, what should we do when someone experiences a highly emotional event? Think about it. If someone has just been involved in a train crash, or witnessed an attack, what should we do? We may feel unclear about how to handle such a situation, wanting to offer support but perhaps worrying about distressing the person further by forcing them to revisit painful memories.

Those who respond to such catastrophes may use something called critical incident stress debriefing to try to help people through this difficult time. It is a process first introduced by emergency health services researcher Jeffrey Mitchell from the University of Maryland in 1983,[26] and it is often referred to as psychological first aid. It is a structured process, administered by trained crisis interventionists. The technique is quite simple, and is founded on the notion that people who have been through an extremely emotional event usually have a need to share their experience with others. In what is known as the recoil phase, people work through what happened to them, and often seek out others who have had similar experiences.

In critical incident stress debriefing people are brought together in small groups, 24 to 72 hours after an event has taken place. Every person attending is encouraged to tell their story of what happened. The intention is to allow people to go from only sharing cold facts, describing the incident with as little detail as possible, to later on exploring their thought processes during the event in detail. It is a gradual and guided disclosure of the event and its repercussions. Participants in the process are also encouraged to focus on their reactions and symptoms, asking questions such as 'What is the very worst thing about this event for you personally?' Finally, participants are educated as to what normal recovery from an event like the one they experienced looks like. It sounds as if it is a well-rounded intervention and clearly it is done with the best of intentions. However, I am afraid to say that I fundamentally disagree with pretty much every part of the process.

It was clearly not formulated by a memory scientist. For one thing, group recollection situations like this are the poster child for people's memories melding into one another's – for better or worse. Due to verbal overshadowing effects, both our own descriptions and the descriptions of others may now become a permanent part of our memory records. Every new account we hear has the potential to taint and re-taint our memories.

I'm not alone in my concern. According to a 2003 review of the academic literature by Grant Devilly and Peter Cotton at the University of Melbourne,[27] the techniques of critical incident stress debriefing can have toxic effects and may even allow for vicarious traumatisation. Vicarious traumatisation occurs when someone tells another person about an event and they experience adverse trauma-like symptoms as a result. Let's imagine that person A and person B were both at an event, but that person A saw gory details that person B was not privy to. In a group session person A talks about these details and explains the terrible effect of seeing them. Later, when person B thinks of the event they think of both their own account and the horrible details told to them by person A. Person B would have been better off had that adverse memory fragment not been imparted to them.

This approach also catastrophises an already delicate situation. Not all people will react to so-called potentially traumatic experiences (PTEs) in the same way. A PTE is an event which is generally considered extremely negative, such as surviving an attack or a natural disaster where your life is in danger. But there is no such thing as an inherently traumatic event – the event only becomes traumatic when the person who experienced it suffers from severe psychological consequences.

The number of people exposed to PTEs varies widely from country to country. According to researcher Dean Kilpatrick from the Medical University of South Carolina and his team,[28] almost 90 per cent of people in the US will experience a PTE at some point in their lives. They found that 8.3 per cent of those exposed to a PTE suffered enough symptoms to be clinically diagnosed with PTSD at some point in their lives. Indeed, when we look at results from various countries around the world, results consistently suggest that only about 1 in 10 of those who experience a PTE suffer severe long-term clinical consequences from it.

So, while some people who experience a PTE will be traumatised and go on to develop post-traumatic stress disorder, others may have almost no emotional response, and yet others may even feel somehow gifted and enhanced by surviving an experience. However, by setting up the expectation that everyone who experienced a particular event probably has – or should have – a severe adverse response, critical incident stress debriefing has the potential to adversely homogenise people's reactions, pushing their memories and responses to be more negative than they naturally would have been.

If a person goes into the debriefing with the view that they have been through a bad situation, but for them personally it wasn't actually too awful, are they really going to say that in the company of people who are tearful and obviously suffering? Instead they may re-evaluate their stance and think to themselves that it *should be* a big deal. They may therefore reframe their original experience and remember the whole situation as worse for them than it actually was, and ultimately have a worse prognosis regarding their ability to deal with it. When we ask people about just how terrible an event must have been, or about how a situation has fundamentally changed them, we are setting up social expectations that they are likely to have a hard time avoiding. We are trying to help but are actually making things worse.

Of course, this kind of sharing can also have tremendously pollutant effects for testimony if the event in question is one of interest to the police. This is because disclosure in such a group setting creates the potential for many of the co-witness effects to be discussed in the next chapter, where memories of events merge and we may appropriate the incorrect details of others, making fertile ground for false memories.

The solution to all this is actually quite simple. If you know someone who has experienced a PTE, make sure they understand that you are available for support whenever they want it. Let them bring the event up if and when they need to, and certainly don't force them to talk about things. They may never want to rehash the experience directly, as they may feel this would re-victimise them, and that is completely okay. Not talking about the event does not mean the person is not coping, or indeed that they are coping, it just means that they don't want to talk. Everyone has their own method of dealing with the aftermath of such events.

From being (or not being) in a helicopter that is attacked, to committing (or not committing) a serious crime, to group disclosure of traumatic events, no memory, no matter how emotional, is safe from corruption. Emotional memories have no special protected place in our brains – they are just like all other memories. Understanding this can make us more considerate of the memory errors of others, can inform our approach to the investigation of criminal offences, and can help us empathise with survivors of extreme situations.

8. SOCIAL ME-DIA

Media multitasking, groupiness, and digital amnesia

Why media moulds our memory

If a tree falls in the woods and no one is around to hear it, does it make a sound? If you have a party and no one Facebooks it, did it really happen? If you have an opinion and you don't tweet it, does it really matter? Deep philosophical questions like this plague generation Y, and media – particularly social media – plays an unprecedented role in people's lives.

Our views on current events have been, and continue to be, influenced profoundly by the internet. And it's not all cat pictures and porn out there – Facebook, Twitter, YouTube, Instagram, Reddit, Upworthy, BuzzFeed . . . we engage with a constant chatter of information which undoubtedly influences the way we perceive the world and shapes the way we share our experiences of it.

Social media enhances our ability to find independent pieces of evidence to validate our memories, but it also has the potential to taint and distort them. We reflect on things that just happened; we document things that we think will get the most upvotes; we filter our lives to look desirable and interesting. But amidst the joy and sense of connectedness all this activity brings us, we occasionally stop and wonder whether this cacophony of impressions is actually good for us. What are the implications of media for our memories?

Media multitasking

Let me tell you a secret. You can't multitask.

While this may not come as a surprise to some of you, many of you probably think that you are excellent at doing multiple things at once. And, who am I to disagree – you are probably more than capable of walking, talking, thinking and drinking, *all at the same time.*

But what we mean by multitasking is generally something more complex, doing meaningful tasks that require attention and memory, and thinking. And, at least since the inception of the smartphone, multitasking seems to have taken on a whole new meaning. We think we can have a conversation over coffee while constantly checking our phones, that we can iMessage all the way through a lecture and still remember the information the lecturer is imparting, and that we can post photos online and simultaneously enjoy the moment.

The basic human assumption that we can adeptly multitask is the result of a fundamental underappreciation of how memory and attention actually work. As neuroscientist Earl Miller from MIT puts it, 'People can't multitask very well, and when people say they can, they're deluding themselves . . . The brain is very good at deluding itself.'[1]

Miller suggests that the better word to use in the sorts of situations that we like to think of as involving multitasking is task-switching: 'When people think they're multitasking, they're actually just switching from one task to another very rapidly. And every time they do, there's a cognitive cost.' So while we may feel like we are getting things done quicker, instead we are just overloading our brains.

A 2014 review of academic research on the impact task-switching has on efficiency, by Derek Crews and Molly Russ from Texas Women's University,[2] suggests that it is bad for our productivity, critical thinking and ability to concentrate, as well as making us more error-prone. And the consequences are not just limited to diminishing our ability to do the task at hand – they also appear to have an impact on our ability to remember things later. Task-switching also seems to increase stress, diminish people's ability to manage a work–life balance, and can have negative social consequences.

In 2012, academic development researcher Reynol Junco from Lock Haven University and sociologist Shelia Cotton from the University of Alabama[3] further examined the impact of task-switching on our ability to learn and remember things, in an article entitled 'No A 4 U'. They asked 1,834 students about their use of technology and unsurprisingly found that most of them spent a significant amount of time using information and communication technologies on a daily basis. More specifically, they found that '51% of respondents reported texting, 33% reported using Facebook, and 21% reported emailing while doing schoolwork somewhat or very frequently'. In terms of time spent trying to multitask while studying, the numbers also added up rather quickly. The students in their sample reported that just while studying outside of class, they spent on average 60 minutes per day on Facebook, 43 minutes per day browsing the internet, and 22 minutes per day on their email. That's over two hours of attempting to multitask while studying *per day.*

Unfortunately for the students, the study also found that such multitasking, particularly the use of Facebook and instant messaging, was significantly negatively correlated with academic performance; the more time students reported spending using these technologies while studying, the worse their grades were. Junco and Cotton concluded that this is possibly due to the students' brains being overloaded, preventing them from engaging in deeper, long-term, learning.

Why exactly does this overload happen? Because, as discussed in Chapter 1, our working memory capacity is incredibly limited, only being able to store four or five pieces of information at once. In 2015 neuroscientists Earl Miller from MIT and his colleague Tim Buschman from Princeton University[4] wrote an article on why we have these limits on our bandwidth of cognition. Every neuron makes electrical noise that can be measured. Brainwaves are essentially our neurons firing together. They can do so at different frequencies, from less than 1 Hz to over 60 Hz. More relaxed states of mind generally create lower frequencies, and the more effort we put into a task generally the higher the frequency goes. These brainwaves are what we can see in some neuroimaging research, like EEG or MEG work. In their study, Miller and Buschman argue that these brainwaves (or, as they call them, oscillatory brain rhythms) are the key to the communication between the neurons in our brain and our core experience of thinking. They suggest that our brain 'regulates the flow of neural traffic via rhythmic synchrony between neurons', meaning that when we have a thought it is because a selection of neurons (which they refer to as an ensemble) are all firing at the same wavelength.

It's like a choir, where each individual member represents an individual neuron. The songs the choir sings are the thoughts in our brain. If each person sings their own song without reference to the others then the result is just a cacophony of sounds. Only when they sing in synchrony do they make coherent songs. Each person can also contribute to multiple songs, but they need to sing differently in order to create those different songs. Finally, the people in the choir don't necessarily sing all the time – they can be part of some songs and not others.

Miller and Buschman argue 'Because ensemble membership would depend on which neurons are oscillating in synchrony at a given moment, ensembles could flexibly form, break apart, and re-form *without changing the physical structure* of the underlying neural network. In other words, this may endow ensembles with a critical feature: flexibility in their construction.' Our brains are able to seamlessly switch from one complex thought to another because neurons can work together by operating on a certain frequency of electrical signal, allowing synchronicity regardless of how they are physically joined together. As the authors put it, neurons *hum* together.

But this ability that enables thought through immediate, temporary communication between neurons also seems to be what makes true multitasking impossible. Our brains can wire and rewire neural networks almost instantaneously, but this mental flexibility comes at the cost of only being able to do one thing at a time. After all, we cannot have the same neurons forming multiple ensembles at the same time, since that would require them to send out different wavelengths simultaneously. The choir members all need to be on the same page.

For example, try looking around the room for things that are both upright and blue. As you do this you are likely searching for things that are upright, and then for each item you identify you switch and ask *blue*? And there is likely a very, very tiny pause as this switch occurs. In an experiment published in 2012,[5] Tim, Earl and their colleagues gave this task to monkeys, training them to switch between paying attention to either the colour of a line or its orientation. The monkeys had electrodes attached to them to monitor their brain activity.

When the monkeys were paying attention, trying to decide

whether a line was red or blue, horizontal or vertical, they produced an increase in a particular type of brainwaves known as beta waves, which fire at 19–40 Hz. Depending on which task the monkeys were currently engaged in – either identifying line colour or orientation – different patterns of neurons were active. Some of the neurons involved in both tasks were the same, but the overall patterns or networks of neurons *humming* together was distinct for each task.

Sometimes the monkey brainwaves would hum at the low frequency of 6–16 Hz – at this level they are called alpha waves. What was fascinating about these alpha waves was that they only seemed to appear when the monkeys were about to switch from identifying the orientation of a line to identifying the colour of it. In other words, the alpha waves were the task-switching waves. Alpha waves help us to stop thinking irrelevant things.

In the monkeys the alpha waves helped to quiet down the humming of the brain's network that was assessing whether a line was vertical or horizontal, so that the brain could hum the line colour identification network instead. This experiment provided hard evidence for the researchers' hypothesis that these two conflicting tasks had to be switched between and could not be completed at once. As such, we cannot hope to make memories for more than one thought at a time.

When we do two tasks that are trying to use the same part of the brain, such as the colour/orientation visual search task, we generally find it far more difficult than two operations that do not directly conflict, such as walking and talking. To look for upright and blue things at exactly the same time (rather than the split-second switch we just described) would require exactly the same visual neurons to do two different tasks at once. If there were little people instead of neurons in your head, this would be the equivalent of you giving Chris two jobs to do in the same minute, and Chris yelling, 'Stop! I need to prioritise one of these!'

On the other hand, we can ask two different parts of the brain to do things at the same time, like giving Chris one job while Adam gets another. They may still slow each other down because they have to talk to each other from time to time, but they can generally get both tasks done pretty well. This is basically what happens when conscious and unconscious processes occur at the same time; Conscious Chris is good at thinking and making decisions, while Automatic Adam is good at driving, walking, and doing other tasks that are mostly automatic to us.

But even this scenario is not so great. Research on the dangers of task-switching shows that trying to divide our attention can be problematic even when it is between seemingly unrelated tasks. In 2006, David Strayer and his research team at the University of Utah[6] published a study comparing drunk drivers to drivers who were talking on their cell phones. In this scenario we can assume that most conscious attention is being directed at the conversation, while driving has been relegated to automatic monitoring. The researchers found 'When drivers were conversing on either a handheld or hands-free cell phone, their braking reactions were delayed and they were involved in more traffic accidents than when they were not conversing on a cell phone.' They went on to say that driving while chatting on the phone can actually be as bad as drunk driving, with both noticeably increasing the risk for car accidents.

The reason for this is most likely that the two tasks, driving and talking, are not as totally unrelated as we might think. This is because Conscious Chris is Automatic Adam's boss. If Adam encounters any situation he cannot solve easily, like having to make a decision, he needs to ask Chris. This is annoying, because it means that Adam keeps interfering with the task that Chris is trying to oversee: *Turn here?* 'Yes, I'll be there for 8.30.' *Can I make it through these lights?* 'I think you should wear the green dress tonight.' Tricky stuff. So even automatic processes are often not as totally automatic as we may assume.

Scientists have therefore been arguing for years that the hazards associated with talking on the phone while driving have more to do with the inability to multitask than the inability to use the hand that is holding the cell phone. The current laws in many countries which allow hands-free phone use, while banning hand-held phone use, seem to be either ignoring or fundamentally not understanding this information.

If I have not yet fully destroyed your view of yourself as a consummate multitasker, I will leave you with one more study that may change your affection towards your phone. In 2015, communications researchers Aimee Miller-Ott from Illinois State University and Lynne Kelly[7] from the University of Hartford looked at how our constant use of our phones while also engaged in other activities can impede our happiness. They argue that we have expectations of how certain social interactions are supposed to look, and if these expectations are violated we have a negative response.

In a qualitative study they asked 51 participants to explain what they expect when 'hanging out' with friends and loved ones, and when going on dates. They found that the mere *presence* of a visible cell phone decreased the satisfaction of time spent together, never mind if the person was constantly using it. Reasons mentioned for disliking the other person being on their cell phone included that it violated the expectation of undivided attention during dates and other intimate moments. When hanging out, this expectation was lessened, so the presence of a cell phone was not perceived to be as negative but was still often considered to diminish the in-person interaction. This corresponded with what they found in their review of the academic literature, where there is strong evidence to show that romantic partners are often annoyed and upset when their partner uses a cell phone during time spent together.

This is also borne out in work by marketing professor James Roberts with Meredith David from Baylor University Hankamer published in 2016.[8] Roberts coined the term 'phub' – an elision of 'phone' and 'snub' – to describe the action of a person choosing to engage with their phone instead of engaging with another person. You might, for example, indignantly say 'Stop phubbing me!' According to Roberts, phone attachment leading to this kind of rude behaviour has been linked with higher stress, anxiety and depression.

So, if you want the interactions in your life to be more productive, safe and meaningful, either be on your phone *or* be present in the offline world, with a definite emphasis on the latter.

Stream of social consciousness

We love our online world because of the constant sense of connection it provides. It grants us access to an almost limitless stream of information about the world, and it provides a forum for us to instantly share our memories and our impressions about it with other people. Through this process of sharing, our memories have become part of a social landscape, a stream of social consciousness that we both shape and are shaped by.

The first time I really appreciated just how much memory can be influenced by social media was in 2011 when I was living in the small city of Kelowna, Canada. Just after 3pm on Sunday 14 August I was out driving with some friends. We turned onto one of the main city streets and felt immediately that something significant had just happened. Kelowna is normally absolutely bursting with tourists in August, and yet this street was eerily empty – no tourists, no locals. Nobody.

While we were looking around, puzzled, a woman ran behind us. She looked terrified. Then suddenly what seemed like every police car in the city came bolting past us. The street was immediately shut down and we were trapped between two police blocks. In an attempt to understand the situation my friend pulled out his smartphone and started to investigate. First, Google – nothing. Then, local news – still nothing. Finally he tried Twitter. Suddenly we had a continuous live stream of what was happening:

'Shots fired.'

'Two gunmen just opened fire on SUV outside Delta Grand hotel.'

'Everybody just hit the ground. Someone just got gunned down outside.'

'Shooters using automatic weapons. The gunmen in a silver van.'

'Medics are removing a man from bullet-riddled car covered in blood.'

'I heard gunshots, it sounds like something is collapsing, like a building is collapsing.'

'War-zone.'

As it turned out, one of the notorious Bacon Brothers had just been shot dead. The brothers were a trio of gangsters who had been implicated in a rash of homicides in the Greater Vancouver area, along with drug production and trafficking. Jonathan Bacon and his family had just been attacked and gunned down in broad daylight by rival gangsters. And the public had documented everything.

We are prone to whip out our phones to film, snap, summarise and post things at the first sign of potential import. Never in history have we had such reliable, independent and ample documentation of important historic events. This ability to corroborate our own appraisals of situations is amazing, but it can also lead to memory conformity – when our way of thinking and our memories become an amalgamated version of the accounts we have seen and heard, and it becomes impossible to distinguish what any one person actually witnessed themselves.

Almost everyone in Kelowna seems to remember the Bacon Brothers shooting the same way. When you talk to people about it their accounts are amazingly, even impossibly, similar. You can probably think of events that you witnessed or were peripherally involved in where the same thing is true. Educational researcher Brian Clark from Western Illinois University argues in his 2013 article cheekily entitled 'From yearbooks to Facebook'[9] that such effects may be because our memory has gone through a transition due to the internet and social media – 'the distinction between public memory and private memory . . . has been blurred to the point of erasure.'

Research has investigated memory conformity in various settings, particularly in eyewitness accounts. In a 2003 paper Fiona Gabbert, Amina Memon and Kevin Allan from the University of Aberdeen looked at how eyewitnesses can influence each other.[10] To test this, they asked two groups of participants to separately watch a video of an event. All participants saw a 90-second video showing a woman going into an empty university office to return a book. Unbeknown to them, there were two different versions of the video, shot from different angles. This meant that participants ended up with one of two sets of information about the event.

The researchers themselves described the difference as follows: 'from perspective "A" (but not perspective B) it is possible to read the title of the book that the girl is carrying, and also observe that she throws a [paper] note into a dustbin when leaving the room. From perspective "B" (but not perspective A) the girl is seen checking the time on her watch, as well as committing an opportunistic crime (sliding a £10 [bank]note out of a wallet and putting it into her own pocket).'

Half of the participants were then asked to work together as teams of two to complete a questionnaire about what happened, while the other half completed the questionnaire alone. Then, after a 45-minute interval, all participants were asked individually about what happened. The researchers found that 71 per cent of those in the co-witness condition reported knowing details that they had obtained through discussing the event with their partners. Further, 60 per cent of those who watched the video from perspective A, where they could not actually see the opportunistic crime taking place, reported that the girl in the video was guilty of a crime. Those in the co-witness group had included on average 21 details that they stole from the other witness. As could be expected, people who completed the questionnaire alone only reported details from the video they had actually seen. Participants in the co-witness group had severely embellished their memory reports with details they had not actually witnessed themselves.

Research like this explores what is referred to as 'post-event information' – information that can influence our memories if we encounter it after we experience or witness an event. It might come from many possible sources – discussing the event with others in person or online, reading articles about the event or related events, seeing photos taken by ourselves or others, to name but a few. Any source of information has the potential to change our memories post hoc.

According to psychological scientist Alan Brown[11] from Southern Methodist University and his colleagues, another source of false memories is memory *borrowing*, where someone directly appropriates someone else's autobiographical memory and relates it as their own. Brown and his team published a paper in 2015 investigating this phenomenon and found that of 447 students who participated in a survey on the topic, 47 per cent answered yes to the question 'Have you ever heard someone's personal experience and later told it to others as if it happened to you?' This means that the students had knowingly claimed authorship of another's autobiographical memory, at least temporarily. Although it may be done consciously, this kind of temporary borrowing can lead to later memory attribution issues, as 27 per cent of the participants also claimed that they had memories which could be their own but could be borrowed from someone else's account of an event – and they were unsure which was the case.

Brown's research also demonstrated that sometimes memory thieves get caught; 53 per cent of participants claimed that they had heard someone tell one of their stories as if it was their own, and 57 per cent claimed that they had disagreed with someone over whether an incident had happened to themselves or to the other person. I personally find this kind of memory thievery particularly applies to family stories, where I occasionally catch myself having to seek confirmation of what really happened from another family member.

So it's clear that memories are contagious; if I let out one of my memories, it is possible for you to catch it and make it your own. And as we merge details from other sources into our own retellings of an event, we have the potential to incorporate both accurate and inaccurate details. In a paper published in 2001, Henry Roediger and colleagues from Washington University coined a good term for this: the social contagion of memory. They showed that one person's memory can be influenced by another's memory errors. A sort of false memory proliferation effect. But why we are so prone to this effect? Researchers argue that it is due to two factors. The first is basic memory distortion; if another person tells you their version of an event, your brain may make new connections that subsequently interfere with your own original memory of it. This is in line with the misinformation and imagination inflation research that we discussed in earlier chapters. The second is source confusion, where we forget the source of information we remember, which can lead us to assume we experienced things that were only told to us.

According to Brown and the other authors of the memory thief experiment, social influences come into play in a number of ways; 'These behaviors appear primarily motivated by a desire to permanently incorporate others' experiences into one's own autobiographical record (appropriation), but other reasons include to temporarily create a more coherent or engaging conversational exchange (social connection), simplify conveying somebody else's interesting experience (convenience), or make oneself look good (status enhancement).' These seem like reasons that are positive and often intentional. But there are scientists who strongly argue for yet another possible social influence: conformity.

Groupiness

The classic studies that first showed our conformity to information when it is provided by others were conducted by psychological scientist Solomon Asch from Swarthmore College in 1956.[12] He found that if we ask people in a group to judge whether or not two lines on a paper are a similar length, their answers will change depending on what others in the group say. Solomon looked at this by planting a number of research confederates who were instructed to give an obviously wrong answer in the room with the participants. The participants thought that all members of the group were other participants, and did not know that they were the only ones who were actually being studied. Indeed, people are often willing to provide a very obviously incorrect answer if it conforms with the answer that everyone else is giving. While we might be comfortable accepting that some people are naturally 'followers' and will inevitably behave that way, what is shocking is that nearly 75 per cent of the participants in the experiments conformed to the obviously incorrect answer provided by the group at least once, demonstrating that actually the majority of us can be influenced by our peers. We can all be victims of situational demands.

When asked later why they had conformed, most participants claimed that they knew the answer was wrong but did not want to stand out. Some, however, claimed that they had really believed that the group must have known the answer better than they did. In 1955 social scientists Morton Deutsch and Harold Gerard from New York University[13] went on to classify social influences of this kind as either normative or informational.

Normative influences are the influences of groups on their members – situations where we do not want to stand out, regardless of whether we believe the group to be correct or not. *Informational* social influences are also facilitated by groups, but do not necessarily require them. They are instances in which we believe that another person is better informed than us, so we adopt their information on the basis that it is probably correct – a situation where a group or, say, an interviewer really does know the correct answer.

These influences help to explain why one person might adopt the account of another. They either might not want to upset the other person by disagreeing (a normative influence), or they may genuinely believe that the other person has a better memory of the account than they do (an informational influence). Of course these social influences are not always a bad thing. If a group of people are running, perhaps they know that there is a fire and you don't – conformity can save lives. It also certainly has benefits for easing conversation and collaborations between members of a group if our memories conform as well. But these social influences become a problem when it is post-event *mis*information that is being spread by them, causing erroneous details to become woven into our memories in ways that can never be untangled.

But that's not all. Deutsch and Gerard coined the term 'groupiness' to describe how cohesive a particular group is – how much its members tend to conform. In sociology the term for this is entitativity, essentially meaning the degree to which the group works as a single entity. We tend to divide the world into 'in-groups' and 'out-groups', meaning those that we identify ourselves as being a part of, and everyone else. For example, your in-group may be your university alma mater, while the out-group may include students from a rival institution.

Dan Ariely, professor of psychology and behavioural economics at Duke University and the author of the bestselling *Predictably Irrational*,[14] argues that our membership of groups makes us exactly that: predictably irrational. Ariely and his colleagues[15] have conducted numerous experiments which have demonstrated that when members of our in-group do something, we are likely to follow suit. This is true for the good and the bad – for example, we are more likely to cheat if at least one of our in-group members does as well. Ariely's research has also demonstrated that we are less likely to conform to those with whom we do not identify – out-group members. This is presumably a deliberate delineation from the behaviour of our rivals – *we are not like them* – as well as an implicit display of solidarity with our in-group brethren – *look how we share values.*

Bearing in mind all these influencing factors, many researchers argue that we need to keep witnesses in police proceedings separate from one another, to avoid tainting effects. And police need to understand that consistency between reports may not necessarily indicate accuracy, only conformity.

Further, the advent of social media has enormously multiplied the potential sources of social influence and misinformation – a friend's Facebook update, a Twitter post by a stranger, a Reddit discussion thread. It seems as though we no longer have full ownership of events in our lives, and are instead living in a time of intense 'transactive memory', as memory researcher Daniel Wegner[16] from the University of Virginia would say. Transactive memories, like our online interactions, are memories that are collectively formed, updated, and perhaps most importantly of all, stored.

Digital amnesia

According to a powerful article entitled 'Google effects on memory' by psychological researcher Betsy Sparrow from Columbia University and her colleagues, 'The Internet has become a primary form of external or transactive memory, where information is stored collectively outside ourselves.'

Sparrow and her team conducted a series of four studies to investigate the consequences of having information constantly at our fingertips. In the first she asked participants to answer a series of tricky trivia questions. She then had them do a word-sorting task that measured the speed at which they classified computer-related words and other words. She found that participants who encountered questions to which they did not know the answers were much faster at sorting computer-related words. She took this to indicate that these participants, when faced with questions to which they did not know the answers, had been thinking about search-engine-related words such as Google and Yahoo. This was taken to indicate that our minds quite automatically wander to search engines as ways of answering informational questions. In other words, when we encounter facts we don't know, we automatically think 'I should Google it.'

The second experiment Sparrow did turned the same trivia questions into statements. For example, she would present the fact 'An ostrich's eye is bigger than its brain.' The participants would then type out the fact on a computer, to make sure they were paying attention. Half of her participants were told that the facts they had typed would be saved for later, while the other half were told that the facts would not be saved. Afterwards, participants were asked to write down as many facts as they could remember. Those who were told that the information would be saved performed worse when their memories were tested than those who did not expect the information to be saved. Sparrow and her colleagues claim that this ties in with the idea that knowing we can always access information later on makes us less likely to put in the effort to try to remember it, therefore decreasing our actual memory for it subsequently.

She repeated the method of study two in study three, except that this time she told participants that the information would either be saved in a specific spot, saved in general, or erased; and instead of having the participants write down the trivia, they did a recognition task. Participants were shown all 30 trivia questions again, but half of them had been slightly altered; the other half were exactly the same as they had seen previously. Participants had to judge whether or not the statements were exactly as they had seen them or not. Again, those who had been told the information they had typed would be erased correctly recognised more of the facts. It seems that we really are less likely to remember information if we think it will be available to us later in a digital form, a phenomenon that is sometimes referred to as digital amnesia. In an age where information is almost always available to us later, this can have profound implications for how we remember it.

Finally, in study four, Sparrow found something particularly peculiar. This time participants were told that all statements would be saved and accessible in one of six folders. For example, participants were told that the fact 'The space shuttle Columbia disintegrated during re-entry over Texas in Feb. 2003' had been saved into a folder named FACTS, DATA, INFO, NAMES, ITEMS or POINTS. When tested on their ability to remember and write down the facts later, it was found that they were more likely to remember *where* the statements were saved rather than the content of the statements themselves. On top of this preference, participants were particularly bad at recalling both the statement and the folder it was saved in. So if they remembered the statement they were unlikely to remember its location, but if they did not remember the statement they were highly likely to remember its location.

It appears that our brains are cognitive misers, picking whichever information appears easier to remember – *either* the statement *or* the location where it could be found again. As Sparrow puts it: 'We are becoming symbiotic with our computer tools, growing into interconnected systems that remember less by knowing information than by knowing where the information can be found.' Take phone numbers as an example. According to internet security company Kapersky Lab, 50 per cent of people cannot remember the phone number of their partner, and 71 per cent cannot remember the phone numbers of their own children – but I bet all of them know where they can find these numbers on their phones.[17]

Outsourcing our information storage in this way presumably means we are potentially even more vulnerable to the kind of post-event misinformation effects mentioned earlier. However, it can free up our cognitive resources to remember other things to which we are less likely to have immediate access elsewhere. We can always look up the name and the fact later, as long as we remember the gist of the information we want to find. Understanding this impact the digital age has had on the way we handle information has the potential to change our approach to education quite radically.

'Perhaps those who teach in any context, be they college professors, doctors or business leaders, will become increasingly focused on imparting greater understanding of ideas and ways of thinking, and less focused on memorisation,' Sparrow conjectures. Being less focused on imparting specific detailed information that students could easily find online, we could instead teach critical thinking so when people inevitably do Google it, they at least know how to find high-quality information and analyse it. Beyond the fact that we seem to encode and remember information differently depending on whether we can access it again later, there are other ways our growing reliance on media can alter the quality of our memories.

You're uglier than you think

Did you know that strangers know what you look like better than you do? Oh, and you also aren't as attractive as you think. I know, it would have been kinder not to tell you. The reason for this is twofold, down to both basic memory processes and the way we use technology.

Let's start with the memory processes involved. Unless you are currently looking into a mirror, your perception of what you look like is a type of memory. It is a memory not only of when you last looked in the mirror earlier today, but of all other times you looked in a mirror or looked at photos of yourself. This means you almost certainly have a composite image of, say, your face in mind when you think of yourself. The problem is that this patchwork memory of what you look like never stood a chance because it can never actually exist in reality. You cannot look today like you did every day until today. Ageing alone makes that impossible, never mind everyday blemishes and style changes. This helps us to understand why, in response to certain photos, we say, 'That's a bad picture of me!' Often what makes it a bad picture is simply that it is at odds with what we *think* we look like, at odds with our memory of ourselves.

In 2008 psychology researchers Nicholas Epley at the University of Chicago and Erin Whitchurch at the University of Virginia[18] published results from a series of studies on how good we are at identifying ourselves, in an article called 'Mirror, mirror on the wall'. They took photographs of their participants and digitally altered them, producing both more and less attractive versions of the original photo by morphing it to fit a standardised highly attractive or highly unattractive face. The photos were morphed to varying degrees, to give a continuum of faces.

Two to four weeks later the researchers presented the various versions, including the original photo, to the participants and asked them to select their unmodified photo out of the series. Most participants selected images that had been morphed with the attractive face to be between 10 and 40 per cent more attractive than the original photo. Participants picked the original, unedited, photo of themselves less than 25 per cent of the time for all conditions. Both men and women clearly thought they were significantly better looking than they actually were, and systematically picked enhanced versions of themselves.

What about identifying others? When participants were asked to identify the faces of friends out of a similar array of modified photos, the researchers found that participants showed a similar bias here as with their own faces – it seems we think that our friends are also more beautiful than they actually are. However, the same is not true for a stranger's face – according to this study we seem pretty good at correctly identifying people we have known for only a short period of time. On average, participants picked photos of themselves that were 13 per cent more attractive than their unedited pictures, picked photos of their friends that were 10 per cent more attractive than their unedited photos, and for strangers picked photos that were 2.3 per cent more attractive than actuality.

We could chalk this bias up to the idea that we generally think we are above average, as discussed in Chapter 6. Or perhaps we could say that because we know ourselves and our friends well, we see an inner beauty reflected outwards. But there is also a third alternative: you have distorted your self-perception, and your perceptions of those closest to you, over time.

This actually has less to do with automatic memory processes and more to do with vanity. We all choose the best photos of ourselves and those closest to us to present to others, paying particular care with photos that we put online or into formal documents. This is precisely the problem – by only including the best photos, with our most dolled-up faces, in photogenic situations, we are setting ourselves up to be bad at recognising ourselves on regular days.

An article published in 2015,[19] spearheaded by psychological scientist David White at the University of New South Wales and supported by the Australian passport office, looked at how good we are compared to strangers at knowing what we look like. In this study the first set of participants were asked to download ten photos of themselves from Facebook, and to rank how much each photo looked like them from best likeness to worst likeness. They were then asked to take a one-minute webcam video of their face and two additional still photos.

These were then used in a face-matching task by a second set of participants who did not know the first set. They were asked to compare the Facebook photos with the webcam video taken during the study and to also rank them on likeness, saying that the photos and videos were either very similar or not similar at all. The strangers picked different 'best likeness' pictures than the participants did themselves. The question became: Who is better at knowing what we look like, strangers or ourselves?

This was addressed by looking at a third set of participants who were asked to match the Facebook photos that had been rated as having a 'good likeness' by the people in the photos themselves, and those picked by the participants involved in part two. This third set of participants was more accurate at matching a person's Facebook photos with the two stills taken in the study when the Facebook photos had been selected by a stranger; photos selected by strangers led to a 7 per cent increase in accuracy in matching photos of the same person. In other words, it turned out that the second set of participants were better at identifying what the first set of participants looked like than the first set of participants were themselves. This might be taken to indicate that strangers know what we look like better than we do ourselves. According to White's team, 'It seems counterintuitive that strangers who saw the photo of someone's face for less than a minute were more reliable at judging likeness. However, although we live with our own face day to day, it appears that knowledge of one's own appearance comes at a cost. Existing memory representations interfere with our ability to choose images that are good representations or faithfully depict our current appearance.'[20]

Research like this seems to indicate that we come to believe that we actually look like the person we present ourselves to be on Facebook and other platforms – we come to internalise our own online facade.

Better together

Our socially constructed memory, biased as it may be, is not *all* smoke and mirrors. Most research in cognitive psychology shows that recalling memories collaboratively is usually disruptive and inhibits accuracy. However, Celia Harris[21] and her team from Australia wanted to change this. In 2011, they published a fascinating article that set out to challenge the existing research methodologies, which often focus on strangers. They wanted to focus instead on people who knew each other very well, and how they recalled shared memories – both of personal and impersonal events.

In the first study of its kind, the researchers interviewed 12 couples who had been married for 26 to 60 years to see how they remembered collaboratively. They asked them to come in for two sessions, two weeks apart. In each interview the participants were given a list of random words to remember. They were also asked some personal details, including names of people they knew. In session one each spouse was asked to recall the impersonal list of random words and the personal list of names on their own. In session two the couple was interviewed together and asked to do the same thing. It was found that some couples displayed collaborative inhibition, meaning that they actually reduced the quality and amount of information that they were each able to remember, while others showed collaborative facilitation, meaning that they helped their husband or wife remember more. Whether partners helped or hindered their spouse's memory depended on *how* they remembered together.

The researchers found that there were both memory-diminishing factors, which were indicated by a lack of cohesion between the partners, and memory-enhancing factors, which were the result of an interactive style of recall. For example, the couples that helped each other remember in an interactive way might have a conversation something like this:

Participant 1: 'What was his name again? Ed something . . .'

Participant 2: 'Yes, Ed Sherman.'

Participant 1: 'So, Ed Sherman was with us at the dinner.'

Participant 2: 'The dinner to celebrate Nancy's birthday.'

Participant 1: 'Yes, with the giant cake that tasted like cardboard.'

In doing so they would fill in each other's memory gaps and work as a team to rebuild the story.

Wanting to go beyond recalling lists of names is work spearheaded by my close friend Annelies Vredeveldt at the Vrije University of Amsterdam.[22] While I cannot say that I am impartial here, I think her research is absolutely fantastic. In one particular study, from 2015, she recruited couples when they were leaving a theatre, having just seen a performance of *Bossen*, a play which includes a three-minute scene in which one of the characters murders his father and then rapes his twin sister. It was the recall of this scene that Annelies and her team decided to focus on. The participants they recruited had no idea that they were going to be asked to participate in a memory study later, so they could not have prepared for this eventuality in any way or watched the play with it in mind.

The couples, who had known each other for an average of 31 years, were interviewed about their memories of the play, first alone and then together. It was found that working together did not necessarily help the participants to remember more about the target violent scene, but they did seem to make fewer errors when working together than when they were interviewed individually. When they were interviewed alone, each spouse made on average 14.6 errors, while when they were interviewed together, each made on average 10 errors. Annelies calls this effect, when we report fewer details that are inaccurate, 'error-pruning'. It may be due to our being more cautious with what we share in company, being less likely to report details we are unsure about. Corroborating Celia Harris's findings from earlier, Annelies also found that certain strategies were better for helping spouses remember: acknowledging what the other person was contributing, repeating and rephrasing their insights, and elaborating on each other's statements.

According to Annelies and her team, 'Taken together, our findings suggest that, under certain circumstances, discussion between witnesses is not such a bad idea after all.' This is good news because most of our memories seem to be in some way collective. According to Elin Skagerberg and Daniel Wright,[23] a whopping 88 per cent of real-world eyewitnesses have co-witnesses, many having more than three others watching an event, and over half discussed the event with at least one of the others who were there. This is much like other events in our lives, where we have at least one friend by our side with whom we immediately discuss what happened.

So, what do we do with these results? We know from previous

chapters that remembering together can have highly problematic repercussions. In situations where memories are shared with others, we can steal them, distort them, or create entirely new complex false memories. Researchers like Vredeveldt agree with such problems of plasticity, but they also argue that under certain circumstances the likelihood of our generating false memories and making errors may not be as dire as in others. In particular, when we know someone really well or use supportive and collaborative memory retrieval strategies we may be at a lower risk for misremembering. However, in practice, the implications of this research are unknown. We haven't studied the benefits of remembering together in nearly enough detail or depth. When working with strangers it's not clear whether we will be able to reproduce these favourable conditions or whether the effects of co-witnessing events and reminiscing about them will always be problematic and lead to memory distortions.

For now, the research strongly suggests putting your virgin memories on paper before they have the chance to be tainted by social processes. After recording a memory in a place that can be accessed later, *then* you can go ahead and share it with others. Just beware that at this point your friends and family can just as easily help or hinder how your memory develops.

A world of witnesses

As soon as we share our lives on our social media sites, we find ourselves involving an almost infinite number of people to be the co-witnesses of our lives. This has irrevocable implications for our memories, for better or worse.

For better, in a very basic sense, remembering life events through social media is going to enhance memories for those particular events. In the scientific literature this is sometimes known as 'retrieval practice',[24] which means that simply recalling information enhances our memory of it. Studies of the effect have shown that simply recalling something can lead to better retention of information than studying the same information for the same amount of time. This line of research suggests that ten minutes of reminiscing may be better for your memory than ten minutes of studying.

Social media also gives us an unprecedented ability to access corroborating evidence for our memories. By Instagramming our food we document where we had lunch and what we ate. By tweeting our opinions we can go back to see if and how we have changed our attitudes over time. By adding friends on Facebook we can see when we first met someone and how our relationship with that person has evolved. We have an astonishing amount of personal data that allows us to track, and confirm, many of our memories. In the case of false memories, this can of course be tremendously useful. With the world as our witness, if we ever get into trouble we can default back to the internet for proof of what happened.

But, from trying to divide our attention more, to having the potential for misinformation to come from virtually anyone, to putting less effort into remembering facts because we can just Google them later, there is also a far more problematic side to social media memory. Having intrusive social media prompts and notifications constantly reminding you of certain events and thrusting more and more information at you also has the potential to severely distort your reality.

This is in part related to the retrieval-induced forgetting effect we discussed in Chapter 3. Every time we remember something, the network of cells that make up that memory becomes active, and that network has the potential to change and lose the details about which we do not directly reminisce. For example, say you are reminded on Facebook of a vacation. The prompt may be a single photo of the event with a caption. As you remember the particular moment in which the photo was taken, it is possible, even likely, that you are forgetting related and unmentioned information of other things that happened during the day.

Of course, it's not just social media that can have this memory-altering effect. Rehashing memories in any situation has the potential to distort them. What is different about social media is that the prompts are being selected from your online persona so they already represent a distorted, social-media appropriate, version of your life. This amounts to a double-distortion – distorting the memory in your brain with a previously distorted memory from your online persona.

By having social media dictate which experiences count as the most meaningful in our lives, it is potentially culling the memories that are considered less shareable. Simultaneously it is reinforcing the memories collectively chosen as the most likeable, potentially making some memories seem more meaningful and memorable than they originally were. Both of these are problematic processes that can distort our personal reality.

How do you know whether you are recalling your experienced reality or your online, crafted, reality? You probably can't tell the difference, as social processes of remembering become magnified and have the potential to infiltrate in ways that were previously not possible. Social media and our ability to connect with others is introducing a fascinating new set of challenges and benefits that memory researchers are only just beginning to explore. It's a brave new world, and we can all look forward to exciting developments in how we remember together.

9. TOOKY PULLED MY PANTS DOWN

Satan, sex, and science

Why we can falsely remember traumatic events

Sometimes false memories can turn into real horrors.

When we find ourselves at the intersections of the concepts we have discussed throughout this book, we can end up in a perfect storm. A storm of misconceived assumptions about how memory works, about faulty memory retrieval techniques, and overconfidence. We can come to spin amazingly complex and fantastical tales of horror that have tremendous implications for our loved ones, for ourselves, and for justice.

To show you what I mean, let me tell you about a case that I consider highly problematic. This story has been recalled many ways by many people, but here is a version that I have pieced together and believe represents the basic outline of what most people would probably agree happened, much of which is supported by a detailed account written in 1995 by investigative journalist Charles Sennott.[1]

The case beings in the spring of 1984. We are in Malden, Massachusetts. A four-and-a-half-year-old boy named Murray Caissie lives here. According to his mother, Murray has been wetting the bed for quite a while, but recently it has been getting worse. That summer he begins to use the same infantile babbling as his 16-month-old brother and acts up more often. He is caught in sexually suggestive play with his little cousin. His mother is getting increasingly concerned about his behaviour.

One evening Murray is having trouble going to pee, and begins to cry. His mother begins to wonder whether maybe his troubled behaviour is a result of something bad that happened to him. This worry begins to escalate sharply, and she apparently wonders whether her son may have even been the victim of sexual abuse. This concern may stem from the knowledge that her brother was abused when he was a young boy.

Murray's mother asks her brother to talk to Murray. Her brother tells Murray the story of the time he was molested when he went away to camp as a child. He says to Murray that he should tell him if anything like that ever happened, if anyone ever did anything to undress him or make him do things that he did not want to do. Murray thinks about this, and then he tells his uncle about being taken to a room on his own where Tooky 'pulled my pants down'. Tooky is the nickname of a man called Gerald Amirault who works at the Fells Acres Day Care Center, which Murray attends. Amirault had once been asked by Murray's teacher to change the boy's clothing after he had wet himself at the Day Care Center several months earlier.

That evening, Sunday 2 September, Murray's mother calls a Department of Social Services hotline and reports that Amirault had taken her son to a secret room and molested him. The Department of Social Services and the Malden police begin to question Murray. They ask him about the event, but the little boy is unable to explain where the abuse happened or what else happened besides Amirault pulling down his pants and touching his penis.

Gerald Amirault is arrested the following day and charged with rape. Over the next week the police go to the day-care centre and request a list of all the children registered there. Is it possible that

Amirault molested other children as well? They interview a number of other children, most of whom, according to the court files, claimed that nothing had ever happened.

At this point, the story hits the media, and parental concern spreads. On 12 September the police call a meeting with the parents of the children enrolled at the centre. Over a hundred parents attend. The parents are informed about the situation. Panic ensues. Social workers are on hand, and distribute a list of behavioural symptoms that indicate sexual abuse. Parents are told to report if their child is experiencing them so that they can be interviewed by a professional in a safe environment. This list of symptoms includes things such as bed-wetting, nightmares, poor appetite and crying on the way to school.

Some of the children meet the criteria. The police give more instructions. The parents are told to ask their children repeatedly and persistently about the abuse. They are told that if the children deny that abuse happened, they should not necessarily believe them. According to some of those who were there, the police explicitly say 'God forbid any of you should show support for the accused. Your children may never forgive you.'

Forty more children are soon identified as victims. Nineteen of them are then interviewed by Susan Kelley, a paediatric nurse who is brought into the investigation because she has published extensively about the horrors of child sexual and satanic abuse. Many of the children initially refute the allegations, but Kelley claims that these children are simply not yet at a point where they are ready to disclose. To create a safe space she uses Bert and Ernie puppets, along with anatomically correct dolls, to talk to and repeatedly encourage the children to disclose the horrors they have suffered.

According to the court files, the children then come to describe in vivid detail the unimaginable ways they were violated. They claim to have had nude swimming parties, and that they were taken to a 'magic room' by a bad clown. According to their statements this 'bad clown' would 'throw fire around the room' while abusing them, and would molest them with his magic wand. They also describe that there was a 'robot like R2D2' from *Star Wars* who would bite them on the arm if they did not go along with sexual requests. They claim to have been molested by lobsters. They claim to have watched animal sacrifices. One of them, a little four-year-old girl, claims that she had a 12-inch butcher's knife inserted into her vagina.

This is only part of the Fells Acres Day Care Center case. There were further accusations. Others were implicated. Many of the children gave testimony and Gerald Amirault was convicted of abuse and sentenced to decades in prison. His mother and sister, Violet and Cheryl, who also worked at the centre, were convicted of similar crimes. All three defendants had denied any wrongdoing. The kind of abuse outlined in the case, if true, is utterly horrific and of course the idea of a victim going unheard after such suffering is appalling. However, cases like this have drawn considerable concern from experts on account of the manner in which the testimony was obtained from the children involved, especially where there is a lack of corroborating evidence.

In 1998 Judge Isaac Borenstein voiced a number of concerns about the Fells Acres case, concerns which led him to overturn Violet and Cheryl's convictions. In his ruling in Commonwealth v. LeFave (1998) Borenstein stated outright: 'Overzealous and inadequately trained investigators, perhaps unaware of the grave dangers of using improper interviewing and investigative techniques, questioned these children and parents in a climate of panic, if not hysteria, creating a highly prejudicial and irreparable set of mistakes. These grave errors led to the testimony of the children being forever tainted.'[2] Gerald Amirault's conviction was not overturned but he was eventually released on parole from the Bay

State Correctional Center in 2004. His mother and sister faced further legal wrangling over their convictions even after their release. Gerald and Cheryl Amirault have maintained their innocence throughout. Violet Amirault died in 1997.

I can't tell you what really happened in this case but I do consider it to be such a confluence of problematic memory-related techniques that I want to pick it apart for you, and hopefully shed some light on how we can avoid such a mess in the future. Let's examine a few critical pieces that build on what we have learned already about the malleability of memory. Consider the pieces of the memory puzzle that are known to allow false memories of traumatic events, such as sexual abuse, to happen. These include a lack of scepticism, assumptions about 'symptoms' of sexual abuse, presumptions of guilt, scientific illiteracy and even, bizarre as it sounds, the presumption of the existence of underground satanic abuse rings.

Sceptics

Let's look at those issues one at a time, beginning with scepticism. Being a sceptic means seeking evidence to support claims, rather than just assuming them to be true. Being a sceptic is different from being a critic, as critics are actively trying to find fault in an argument, while sceptics are looking for evidence both for and against a claim. What we can be sure of in the Fells Acre case is that at the time of the allegations most of those conducting the investigation were not sceptics.

One of the teachers who used to work at the centre, and was interviewed by journalist Charles Sennott, claimed 'They scared the hell out of me . . . The whole thing was geared toward convicting them, no one was asking, "Do you think this really happened?" No one wanted to hear anything that was common sense.' Rather than scepticism about the fantastical stories of clowns, robots and lobsters molesting the children, the interviewers assumed that these were reasonable descriptions by young children who did not understand their experiences. A burden of proof was dismissed in favour of an explanation that fit the desire to catch predators that were assumed to exist.

In this case, there really was a shocking lack of evidence. Some of the allegations, including the horrific account involving the butcher's knife, should have left wounds or scars. Yet there were no scars or suspicious injuries on any of the children. It would have also been reasonable to assume that if such heinous atrocities were repeatedly occurring, some of the other teachers might have noticed something amiss. But no confirmatory accounts were provided by any other teachers. The supposed 'magic room' the children described was never identified. When the videos of the interviews by the police and the therapist were reviewed later, investigators could see that the children almost always denied the abuse initially, but were encouraged with toys and led by suggestive interviewing until they gave the answer that seemed to suggest abuse.

According to psychologist Maggie Bruck from Johns Hopkins University, who has conducted extensive research on interview tactics that can lead to children fabricating stories, 'It is beyond me to tell you whether these children in the Amirault case were telling the truth or not. But I can tell you the questioning was a highly suggestive interrogation process and in that context, our research shows, children fabricate stories.' If we apply what we learned in previous chapters, we can see how the investigative techniques used here were leading and suggestive, and were combined with the kinds of imagination exercises that are known to create false memories.

These issues were ignored, probably because the accounts given by the children were detailed and emotional. Of course when a case involves allegations of such appalling abuse it is important to give credence to potential victims, but to have blind faith in memory accounts is ill-advised. Even vocal critics of the case who believe that the abuse never happened do not say that the children were lying, just that they had their memories misled. As we know from the literature on false memories that has been discussed throughout this book, false memories often look and feel real, and since they are not a form of lying, are unlikely to look deceptive.

So, problematic interview techniques were used which call the accusations into question. This is easy to judge through the lens of hindsight, when errors of non-sceptical thinking may be easy to see. However, if our children, friends, or family came to us with allegations of gross indecency, would we not react in the same way? Allegations would probably lead to assumptions; assumptions would lead to visceral responses. 'Catch the monster!' we would likely cry. We would not want to rest until the culprit was found and put away. We would likely pay little credence to the possibility that our loved one could unintentionally generate a rich false memory of such a traumatic event. Never mind a memory so incredibly detailed.

Built into this case, and many others, is the assumption that we can tell, based on someone's behaviour or emotions, that they have been sexually abused. But can we?

Sexual abuse accommodation syndrome

In the Fells Acres case, the little boy Murray was judged to be showing signs of having been sexually abused. These signs included bad behaviour, sexualised play, and bed-wetting. The assumption of behavioural symptoms of abuse was further endorsed when the other parents at the day-care centre were given a list of symptoms regarding what to watch out for in their children. According to one of the parents 'They had a list in the newspaper, what you look for when a child has been sexually abused . . . The sleepless nights, the nightmares, the bedwetting.'

All this is not surprising, given that in 1983 there were some important changes in legislation regarding the reporting of child sexual abuse in the US, and in the same year the 'sexual abuse accommodation syndrome' was proposed.

Sexual abuse accommodation syndrome was a concept first published by medical doctor Roland Summit.[3] Summit wanted to create a model for use in light of the increased awareness and concerns regarding sexual abuse at the time. He had a number of ideas, including some directly related to the disclosure of sexual abuse by children in interviews. He claimed that because of the traumatic effects of sexual abuse, children would suffer a number of psychological consequences, including shame, embarassment, allegiance to the perpetrator, and a sense of responsibility for the event. Because of these psychological responses, Summit claimed that children would often delay the disclosure of abuse, deny the abuse, and recant their allegations. According to psychological scientist Kamala London from the University of Toledo and her colleagues in a study published in 2008,[4] 'Summit's 1983 paper has exerted a tremendous influence on forensic interview practices'.

The paper was particularly influential for therapists like those involved in the Fells Acres case. This meant that the initial denial of sexual abuse by almost all the children in the case was seen as a psychological defence mechanism that was to be ignored, and the assumption of abuse was sustained despite a lack of allegations. Renee Brandt, the lead therapist who assisted the children's interviews, said in court during the trial against Gerald Amirault that it was normal for a child not to admit abuse right away, and that 'a child begins to disclose the sexual abuse and there is less repression of this material in their mind. So that it is very often the case that symptoms will either appear for the first time or, even if they were present, become quite exaggerated'.[5]

So, what is the evidence for this model, which supports the idea of this abuse denial? Kamala London and her colleagues reviewed how children disclose sexual abuse to others. They summarised the evidence across dozens of studies and found that 'among valid abuse cases undergoing forensic evaluation, denial and recantation are not common'. They claim that the scientific literature does not support the model proposed by Summit, because they found a 'dearth of empirical support'. In other words, the idea that children often deny abuse when confronted about it is largely a myth.

Further, London and her team claim that the kinds of symptoms that are often assumed to be indicative of child sexual abuse are not useful because 'psychological and medical profiles do not reliably differentiate abused and non-abused children'. They specifically say that the kinds of behaviours that may be related to abuse, such as anxiety, bed-wetting, and sexualised play, 'are also present in many non-abused children'.

Yet, the assumption that there are reliable symptoms of abuse continues to be persuasive. For example, the 2016 website of the National Society for the Prevention of Cruelty to Children in the UK has a list of symptoms supposedly indicative of abuse. The list includes things such as 'Acts out excessive violence with other children', 'Poor school attendance and punctuality', and 'Wets or soils the bed'.[6] It is of course true that these symptoms *may* be associated with abuse but it is important to emphasise that it is also perfectly possible that they are not.

A 1993 study by psychological scientist Kathleen Kendall-Tackett[7] and her colleagues from the University of New Hampshire synthesised the research on individual types of risk factors, and provided clear evidence that after experiencing sexual abuse, children do indeed display certain symptoms. For example, on average 33 per cent of those who had experienced sexual abuse displayed some sort of fear, 53 per cent showed general symptoms of PTSD, and 28 per cent showed inappropriate sexual behaviour.

However, critically, approximately a third of the children across the various studies they examined showed no symptoms at all. This means that while many children may display symptoms often associated with abuse, there are no symptoms that are universal, and many children don't meet this framework at all. According to the authors, once again, 'The findings suggest the absence of any specific "sexually-abused-child syndrome" and no single traumatizing process.' In other words, we should not ever make an assumption that a child was abused simply because of their behaviour, there are simply too many other reasons that have nothing to do with sexual abuse that can explain why a particular child may be disruptive, violent, play truant from school, or wet the bed. Assuming that these are reliable symptoms can heavily influence the way interviews are conducted, potentially leading to the police using – albeit perhaps unintentionally – suggestive and leading questioning techniques, in an attempt to make the child disclose the abuse.

Another idea that fed into the Fells Acres case, as well as many others, is that of a hidden world of satanic ritual abuse. This theory was incredibly prevalent at the time, and continues to be relevant in some parts of the world today. Let us explore this next.

Satanic panic

More than once have problematic lines of legal inquiry led to very specific accusations of satanic sexual abuse perpetrated by childminders against young children. A whole slew of accusations related to paedophilic sex rings took shape particularly during what became known as the 'satanic panic'– the day-care sex abuse hysteria of the 1980s and 1990s that ended up affecting parents and police all over the world. Few things have contributed so much to the satanic abuse hype as the bestselling book *Michelle Remembers* by Michelle Smith and Dr Lawrence Pazder.[8]

The book was a smash hit when it came out in 1980. It chronicles the real-life therapy sessions which took place between Pazder, a psychiatrist, and Smith, his patient. Smith apparently had begun to see Pazder in 1973 in his private practice in Victoria, Canada. During the course of her regular visits – Smith was being treated for post-miscarriage depression – she told Pazder that she felt she had something important to disclose but was unable to remember it. This strange forgotten event seemed important to the good doctor, so much so that over the next 14 months Pazder allegedly dedicated over 600 hours to trying to help Michelle remember her forgotten past using hypnosis.

And remember she did. In one of her therapy sessions she apparently screamed for 25 minutes and started to speak in the voice of a five-year-old. She began to recall abuse, satanic ritual abuse, perpetrated by her late mother and others who she claimed were part of a satanic cult in Victoria. She remembered that as a five-year-old she was tortured, sexually assaulted, locked in cages and forced to partake in horrendous rites, and witnessed ritualistic murders. She even remembered being covered in blood and being rubbed with the dismembered body parts of babies and adults who had been sacrificed.

When the book came out it caused a sensation and started a public discussion about the notion of accessing long-forgotten memories in therapy – particularly of ritualistic sexual abuse. It also started a conversation about the alleged increasing problem of satanism and this particular kind of abuse. Following its release there was a surge in alleged victims of satanic ritual abuse coming forward, and lawyers who were preparing their cases against the alleged satanists in such cases used the book as an informative resource. It was even used as training material for social workers. Pazder went on to become a key figure in psychology, an expert in the field of this growing issue. The book was widely considered important, timely and accurate.

Unfortunately – or perhaps fortunately, given the lurid claims it makes – it seems that the whole story was bogus. The book has since been repeatedly discredited by journalists and investigators, who found no corroborating evidence for the allegations it contains. Surviving family members refuted those allegations. Many of the events that supposedly occurred are incredibly unlikely or even impossible, and inconsistencies abound. The account which set an entire generation of allegations into motion seems likely to have been based on a writer's imagination.

One of the leading researchers on false memory, Elizabeth Loftus, wrote a number of review articles in response to this bout of sexual abuse hysteria. In one classic article, entitled 'Who abused Jane Doe?',[9] along with her colleague Melvin Guyer, Loftus argues that there is no science to support the therapeutic techniques, including hypnosis and regression, used in these kinds of cases. She claims that the evidence that these therapists, and proponents of recovered memory therapy, rely on is highly questionable, because there is no evidence that repressed memories exist in the first place. This is something we will explore shortly.

Loftus concludes by essentially saying that these techniques are quackery, and that what ends up being counted as evidence in trials can be the result of a cascading effect of assumptions leading to the generation or misinterpretation of evidence. In order to understand this more clearly we need a basic introduction to Freud.

Sex with Freud

We have Sigmund Freud to thank for the whole idea of 'repressed' memories such as those outlined in *Michelle Remembers*. Freud, the Austrian psychiatrist extraordinaire. Freud, the man who revolutionised our understanding of psychology. Freud, the man who gave us the conscious and subconscious. The id, the ego and the superego. Freud, repeated nominee for, but never winner of, the Nobel Prize. Freud, the psychoanalyst. Freud, who lived around the corner from me.

Sigmund Freud spent his last year of life in Hampstead, London. The house where he lived, which has since been turned into the Freud Museum, is a beautiful red-brick build with white trim on a leafy green street. A friend of mine calls this style of architecture 'gingerbread house', and the neighbourhood really does have a slight Grimm's fairy tale feel to it. It's also my neighbourhood; I live a ten-minute walk from the house. Unfortunately the last year of Freud's life saw the degeneration of both his mental and physical health, so he likely spent most of his time in a wheelchair at home. Despite knowing this, when I walk around my local area I sometimes picture Freud walking with me. An impossible but delightful image.

Except, if I were to actually meet Freud today, we would almost certainly hate each other. We would have major epistemological disputes. To understand why I feel this way, you need to understand Freud's contribution to the current discourse on memory, and where Freud went wrong on this issue. Conveniently, in 1995 critical author Richard Webster wrote a book entitled *Why Freud Was Wrong*,[10] in which he describes Freud's concept of psychoanalysis as perhaps the most complex and successful pseudoscientific theory in history.

In essence, Freud made a number of problematic assumptions.

First, he assumed that there is an unconscious mind which can store memories, emotions, desires and motivations which are unacceptable or unpleasant and which are therefore actively suppressed by the conscious mind. According to Freud, even though we are generally unable to directly access the unconscious mind and the traumatic memories that it hides from us, it still manifests in our lives through our behaviour.

This is akin to the assumptions discussed earlier about symptoms such as bed-wetting, which were also presumed to be the behavioural consequences of deep psychological trauma. Freud based most of his published theories exclusively on interviews with his patients, not on science. Indeed, it is tempting to say that Freud was not a scientist at all. If you don't believe me, ask the Nobel Prize committee. After 12 years of Freud being nominated for the Nobel Prize, the committee actually hired an expert to inquire into his work. The expert came to the conclusion that 'Freud's work was of no proven scientific value'.[11]

Freud's second assumption was that many physical and psychological disorders, if not all, are the result of trauma in childhood. Much of the foundation for this view came about when Freud formulated theoretical speculations in his practice as a physician regarding a disease he called hysteria, which he defined as a mental unrest. This mental unrest, he claimed, was the result of internal psychological conflict leading to physical symptoms, symptoms which could include amnesia. This supposed condition was thought to almost exclusively affect women – years later feminists understandably tore the idea to shreds.

The most common cause of this disease, Freud claimed, was repressed sexual abuse. This meant that many of the women who sought Freud's treatment were automatically assumed to have experienced sexually traumatic childhood experiences. If they refuted this idea, Freud took this as proof that such abuse had occurred. As he says in his own records: 'Before they come for analysis, the patients know nothing about these scenes.'

The third assumption that Freud made was that all of this could be treated in therapy, mainly through the use of imaginary reproduction. Freud would have his patients picture the sexual abuse that he assumed they had experienced, ignoring protests that such events had not occurred. He encouraged them to picture events in as much detail, and for as many therapy sessions, as possible. He thought this technique, also called regression, would allow them to access their repressed memories.

Freud thought these processes were forcibly accessing the subconscious, rather than simply forcing the creative parts of the brain to generate horrible fictions. According to his own accounts patients 'are indignant as a rule if we warn them that such scenes are going to emerge. Only the strongest compulsion of the treatment can induce them to embark on a reproduction of them.' He believed that only through his regression technique could he make trauma buried in the unconscious move to the conscious, where it could be worked through and dealt with in therapy.

False memory researchers like Chris French, however, argue that the very foundations of these assumptions are inaccurate. The idea of conscious memories being separate from, and in conflict with, unconscious memories was never based on science. In 2015, after decades of research on memory in police investigations, French stated unequivocally that 'there is no credible evidence for the operation of this psychoanalytic notion of repression and very strong evidence that the conditions under which therapy takes place are indeed ideal conditions for the generation of false memories'.[12]

And false memory experts Stephen Lindsay and Don Read state that 'extreme forms of memory work in psychotherapy combine virtually all of the factors that have been shown to increase the likelihood of illusory memories or beliefs'. They suggest that this is because of four problematic situations that are commonly present when someone wants to access what they think is a repressed memory of trauma.

The first problem is that an expert, often a therapist, proposes the idea of repressed memories to their client. They may say something like: 'Many people push bad memories into their subconscious, and this can have lasting repercussions for our mental health.' This idea may be substantiated by the expert claiming that the patient shows symptoms of repression like those discussed earlier, such as anxiety or depression. 'You know, anxiety is a common symptom of a history of trauma.'

Second, the expert tells the patient that they need to uncover the repressed memory in order to heal their symptoms.

Third, the patient is then given suggestive and leading information from books, anecdotes, or the therapist themselves.

Fourth, the details of a basic trauma are often given to the patient, and the patient is told to visualise them in line with a memory script. 'Just picture a trauma happening, and the memory should start to come back to you.'

We can see the similarities between this approach and the potential problems which can occur in modern abuse investigations. As we have seen, current research (including my own) has shown that these are exactly the kinds of conditions which can foster the creation of false memories. And, unfortunately, although they have all been discredited, Freud's assumptions about memory repression, the subconscious and retrieval therapy are all still represented in a subset of the therapeutic population.

According to a comprehensive survey published in 2014 by memory scientist Lawrence Patihis from the University of California, Irvine, and his colleagues,[13] misconceptions about repressed memories have declined since the satanic panic, but certainly still exist. In their large international sample, 6.9 per cent of clinical practitioners believed that 'many traumatic memories are often repressed', as did 9.9 per cent of psychoanalysts and 28 per cent of hypnotherapists.

Where there is smoke there is fire

Another problematic attitude that we see influencing cases involving alleged abuse is the logical fallacy that 'Where there is smoke there is fire.' I shudder at the blithe certainty contained in that statement every time I hear it. I cannot help but wonder at the mental gymnastics the person in front of me must be doing to reconcile such a view with modern notions of justice. They are twisting innocent until proven guilty into guilty until proven innocent; the assumption of that statement clearly being that when individuals are accused of a crime they are probably guilty. Even when a person is exonerated, popular notions that the alleged crimes must have occurred often persist. Even if no evidence is found, no scars are revealed, and alibis are solid, accusations can override our better sense of justice.

The whole process of pursuing allegations of ritualistic child sexual abuse, especially following questionable interviewing or therapeutic interventions, has been likened to the Salem witch trials, mostly because of the non-sceptical acceptance of the allegations involved, and the absurd burdens of proof which are placed upon the accused.

During the infamous Salem witch trials, which occurred when in Massachusetts between 1692 and 1693, over 200 people were accused of practising witchcraft, some of whom were executed. Accused individuals were given trials such as the swimming test, where they were thrown into a nearby body of water to see whether they would sink or float. It was believed that witches would float, since the water would reject them, while an innocent person would sink. Needless to say, many individuals drowned in the process, and those who did not were labelled witches.

Similarly, accused witches were at times stripped and their bodies inspected for 'the Devil's mark' meaning a skin tag, birthmark or other blemish which the Devil had placed upon their body. Because such marks were conveniently thought to be amenable to change in shape and type, virtually any imperfection could be taken as proof that an individual was a witch. And of course there was no escaping such charges – how do you prove that a particular mark was *not* created by the Devil?

This is not dissimilar to the treatment of many individuals accused of satanic child abuse, imprisoned despite the lack of corroborating evidence, and with an impossible burden of proof upon them to demonstrate their innocence – it can be extremely difficult to prove that something did not happen if the only evidence is a verbal account and the event is said to have taken place in private.

Having said all this, some experts have disagreed with the witch-hunt narrative of abuse cases. One such person is Ross Cheit, Professor of Political Science at Brown University, who previously obtained a law degree and a PhD in public policy from UC-Berkeley. In 2014 he published his book *The Witch-Hunt Narrative: Politics, Psychology, and the Sexual Abuse of Children*[14] in which he shares the conclusion of 15 years of his own research, during which he dissected old trial transcripts and interview tapes, examining cases of alleged child abuse.

Cheit claims that the witch-hunt narrative represents a fundamental misunderstanding of the actual prevalence of child sexual abuse. He reports that child sexual abuse is actually a lot more common than many people would like to think. He claims that those who have sought to diminish that idea have tended to pull out individual problematic cases – and these undoubtedly exist – and use them to discredit the idea of widespread sexual abuse. As the back of his book states, 'purveyors of the witch hunt narrative never did the hard work of examining court records in the many cases that reached the courts throughout the nation. Instead, they treated a couple of cases as representative and concluded that the issue was blown far out of proportion.' He is claiming that false memories of sexual abuse can occur, and that there are indeed problematic cases, but that this is uncommon and can paint a distorted picture of the many cases involving claims of events that actually do take place. Cheit continues to be vocal about what he considers to be an overstatement of the prevalence of false memories of sexual assault. He believes that the entire rhetoric of scepticism is problematic, for real victims and the justice system.

It's certainly an extremely complex issue, and the concerns that Cheit raises are completely understandable. No social scientist would ever argue against the notion that child sexual abuse is an incredibly important issue, or against the idea that most people who approach the police with historical abuse cases are accurate. Most cases of sexual abuse *are* valid, there is gross underreporting of it, and we desperately need victims' voices to be heard. While Cheit may suggest that those who study false memories are peddling notions such as 'the belief that the charge of child sex abuse was typically a hoax', no psychologist or scientist I know would ever argue such a horrible thing.

What we do argue is that suggestive and leading interview techniques can lead to the false recall of terrible things. We also argue that pursuing child sexual abuse allegations, especially when absurd event details are involved and no corroborating evidence can be found, needs to be approached with extreme caution. Caution because false memories of traumatic events clearly exist, because they seem incredibly real, and because our reaction to such allegations is often led by visceral responses rather than rational ones. In seeking justice we must remember that as well as protecting the victims of abuse we must do our best to protect the falsely accused, and that means that statements must not be gathered using methods which could potentially seed false memories.

False memory 'syndrome'

The final factor which can cause problems in criminal cases is scientific ignorance. Many of the professionals involved in such cases are not aware of (or trained in) what the latest research says about memory.

For one thing, I often encounter the use of the term false memory *syndrome* by lawyers, therapists and the police. This term is simply inaccurate, false memory syndrome does not exist. The use of the word 'syndrome' has an inherently medical connotation, almost as though one could catch a false memory like one can catch a cold. It also has the connotation of being an abnormal process. But, as we know from the research covered by this book, such a conceptualisation is simply not true. We are all capable of forming elaborate false memories, and small false memories happen all the time without our knowledge. False memories are just memory illusions due to normal kinds of memory processes. Thus, the correct thing to do is simply to say that someone has – or may have – a false memory, omitting the unnecessary term 'syndrome'.

Psychologist Michelle Hebl from Rice University and her colleagues summarised the current stance well in 2001:[15] 'Although the terminology implies scientific endorsement, false memory syndrome is not currently an accepted diagnostic label . . . this syndrome is a non-psychological term originated by a private foundation whose stated purpose is to support accused parents. Terminology [which] implies acceptance of this pseudodiagnostic label may leave readers with the mistaken impression that false memory syndrome is a bona fide clinical disorder supported by concomitant empirical evidence.' Scientists today do not use the term false memory syndrome.

By now, if you have read most of the chapters that make up this book, you probably have some appreciation of the incredibly varied things that can cause false memories – from our basic brain biochemistry, through our tendencies towards overconfidence, to faulty interview tactics that can generate complex fictitious accounts.

However it is unfortunately the case that, even after decades of evidence piling high, some people still remain unconvinced that rich false memories of physical and sexual abuse can exist. Particularly in the mid-1990s, the field of false memory research came under heavy attack. Advocates of the existence of false memories were accused of saying that victim statements, if they had been made in questionable circumstances, were therefore likely to be false. This is certainly not the general stance taken by researchers – not only would that be offensive, but also counterproductive for actual victims of abuse.

Nevertheless, battle lines were drawn and the so-called memory wars began, leading to an unbelievable crossfire of academic research, lawsuits, and sensationalised news articles. Opponents of false memory research often claim that we are silencing victims and defending the guilty. There is, of course, a legitimate concern here – it would be an awful thing for someone who has had any kind of traumatic experience to be disbelieved. But given that there is empirical evidence that false memories do exist – and can be created – any conception of justice must surely also be concerned with trying to protect the innocent from false conviction. There is no doubt that this is a highly sensitive, difficult area, but sweeping the idea of false memory away completely and trying to pretend it does not exist is unhelpful. Ultimately there is no quick resolution.

Hired guns

If you tell people that you professionally question all memories, and are on balance more likely to serve on the side of the defence than that of the alleged victim in a trial, you are seen as an advocate for criminals. People ask me all the time, 'How do you know that you aren't helping rapists and murderers get away with their crimes?' The answer is that I do not. I am certain that some guilty perpetrators have used accusations of false memories as a successful defence. The same must surely be true of other legal defences – someone might successfully make a plea of temporary insanity when they committed a crime in full knowledge. However much we may deplore that, it does not diminish the importance and existence of legitimate cases of temporary insanity.

So while recognition of the existence of false memory and other memory distortions has this potential cost, I fully believe that my work, and the work of my colleagues, helps to uphold the course of justice overall. Everyone has the right to a fair trial, and that trial is only fair if there are empirically based standards of evidence. We need to ensure that trials become less biased against people who are accused of crimes, and take more to heart the truism that just because you have been accused of doing something bad it does not mean you did it. Given what I know about memory,

I would not want to live in a world where a single memory by itself is enough to enact legal sanctions.

According to the international organisation the Innocence Project,[16] which is dedicated to exonerating convicts who they strongly believe are innocent, faulty memories, particularly of eyewitnesses, are the major contributing factor for wrongful convictions. For example, in 2015, out of 325 cases where modern DNA testing proved innocence beyond reasonable doubt, a whopping 235 cases involved eyewitness misidentification. This suggests that false memories play an absolutely critical role in the imprisonment of the innocent.

There are no easy answers here, but let's not deny the reality of false memories. Let's tell everyone that they exist, that they can look and feel like real memories, and that we can even misremember highly emotional or traumatic events. Let's increase insight into how our beautiful minds work, accept the malleability of memory as a part of life. Knowledge is power, and ultimately greater knowledge of this subject empowers us all and protects us from interview techniques and assumptions that can spiral out of control.

Rich false memories exist, whether we want them to or not.

10. MIND GAMES

Secret agents, memory palaces, and magical realism

Why we should embrace our faulty memory

If I have done my job, your memory should now seem hopelessly fragile, impossibly inaccurate. To bring you to an acceptance that all of us have critically flawed memories is the very reason I wrote this book. You hopefully now appreciate just how plagued memory is by biological flaws, perceptual errors, contamination, attentional biases, overconfidence and confabulation. But where does this leave us? We cannot possibly just write off memory as a lost cause. We still need it. We rely on it, every single day of our lives.

As I have briefly mentioned once or twice already, metamemory is our knowledge of our own memory and the way in which it functions. This is a type of metacognition, a type of thinking about thinking. Having this ability means that we can muse about why we remember, how we remember, and how good we are at remembering individual pieces of information. One of the first experimental studies on metamemory was conducted in 1965 by Joseph Hart.[1] He wanted to understand a particular feature of metamemory, a construct he called the 'feeling of knowing'.

When you *know* you know

Hart described the feeling of knowing as the sense we have when we think we have something stored in our memory but cannot recall it. He wanted to know whether this feeling is associated with accuracy; whether when we feel like a memory is there but we cannot immediately access it, we are correct in assuming that the memory exists. Over many years of research, he demonstrated that when his participants had this feeling of knowing, they were often correct. However, this information could only be recognised, not recalled. This means that, for example, if we have a feeling of knowing, we are likely to be able to correctly recognise information in a multiple-choice test but will not be able to produce it from scratch in an open-ended question. For open-ended questions we need more than just a feeling of knowing, we need memories that are more accessible.

In a more recent study of this phenomenon from 2014, psychological scientist Deborah Eakin[2] and her colleagues at Mississippi State University wanted to know whether the feeling of knowing was generally accurate regardless of age. They had students, who were on average 19 years old, and seniors, who were on average 72 years old, complete a memory experiment. Participants came into a lab twice to complete a computer task. The first time they came in they were shown a series of pictures of faces. Some of these faces were of famous people, and some were faces of unknown people. This acted as a foundation for the researchers to filter out faces that the participants already knew.

Participants were then asked to come back a week later. Here they were given the faces of non-famous people again, but this time they were presented with names. The participants were instructed to remember the name–face pairings. After being shown all of the names and faces, the participants were asked how good they would be on a multiple-choice test at recognising the name for each face, from 1 to 100. They were then actually tested, being asked to recognise the name of the face they were shown from a set of three.

As the researchers had predicted, the university students were better at remembering new faces and names than older participants, which is in line with a general decline in ability to learn new information as we age. However, they found that the feeling of knowing was the same for all participants. Both younger and older participants generally accurately predicted when they would know information.

This lends support to the idea that we may have an intuitive understanding of the things that we remember and things that we do not. It is the reason why we sometimes say things like 'I'll know it when I see it' – because sometimes we just *know* we know something, even if we can't directly remember the information.

But, wait. Doesn't this fly in the face of many of the things I have said in this book? I have repeatedly demonstrated that we are overconfident in our memories and really cannot rely on our instincts as a guide regarding memory veracity or accuracy. Am I suddenly backtracking? No, I'm not. While our feelings and thoughts regarding whether we have stored a memory are often linked to that information being there, they also often are not.

If we take Eakin's face–name matching task as an example, younger participants rated their feeling of knowing for the multiple-choice test as 42 out of 100 for items they later recognised, and 24 out of 100 for those they did not later recognise. In other words, even for information that they could *not* later remember participants still had a pretty high feeling that they would be able to anyway, an error which is akin to our overconfidence bias from Chapter 6. This shows that this feeling is still far from perfect and can mislead us. A classic feeling of knowing error in everyday life is a situation in which you think 'I know that guy' because his face seems familiar, even when in reality you have never seen him before.

When we start questioning our metamemory insights, like wondering where our feeling of knowing is coming from, we begin to talk about metacognitive monitoring of metamemory. *Meta*-metamemory. It is at this point that we transcend the regular human pursuit of thinking about how good our memory is, and replace it with a much larger scale question of *why* we think the way we do about memory. Within this meta-metamemory there is a great deal of frustration, as we come to wonder whether any of our memories are ever trustworthy, but also a great deal of potential for maximising our memory ability. And actually, although I have been at pains to point out the many ways that our memories are faulty and can lead us astray, I must also emphasise how incredible they are – to have a biological system which associatively stores such vast amounts of information is a miracle of evolution and we should all consider ourselves extremely lucky.

Let us explore some ways in which we can harness our faulty memories to work to our advantage, and to potentially change how we reflect on our own lives.

Brain games

Some of us aspire to know everything – or at least to know more. And in the quest to make us cleverer and more knowledgeable we seek out memory training. There is now a huge and growing industry in smartphone and computer games that promise to make you smarter and to improve your memory. The ads say sexy things like 'scientifically validated', or that they will give you 'bespoke brain fitness', or claim to be designed by neuroscientists.

They usually say that their benefit is proven because you get better at the games as you play them. 'Look! When you first started you could solve a Sudoku in ten minutes, now you can do it in two minutes!'

Following this argument along we could potentially argue that *all* games train our brains, since we almost always get better at them with practice. And throwing in a claim that a game can physically change your brain is extra manipulative, as this is of course always true. Technically everything we do changes the physical structure of our brain, if ever so slightly, so of course playing brain training games will also do this.

These kinds of games also tend to promise 'transferable' gains: brain improvement that can be learned in a game but used for tasks in other contexts. Many of them claim that they focus on building our fluid intelligence, which has to do with our capacity to think logically and solve problems in novel situations. Fluid intelligence is dependent on our working memory – how much we can keep in mind at any one time. Naturally the idea that we could play a game to increase our working memory and raise our IQ accordingly is an alluring one.

One way to measure fluid intelligence is via Raven's Progressive Matrices test, often referred to simply as the Raven test. This involves completing sets of patterns that get harder as the test progresses. These puzzles are purely visual, and usually involve basic shapes like squares and triangles. The test is supposed to measure two types of fluid intelligence: the ability to make sense of complexity and the ability to store and reproduce information.

In 2008 two researchers set out to test whether memory games can increase performance on the Raven test. Psychological scientist Susanne Jaeggi and her colleagues at the University of Bern[3] wanted to see what would happen if they had participants play a brain game called the *n*-back, which involves being presented with a series of stimuli – typically pictures, numbers, or letters – each spaced several seconds apart. So you might get L . . . M . . . K . . . M. The participant then has the job of deciding whether the current stimulus (a letter in our example) matches the one that was displayed *n* trials ago, *n* being a stand-in for the number of places in the series the participant has been instructed to go back. For example if you are doing a 2-back, you want the participant to say whether the letter in front of them now is the same as the one they were shown two letters back. The more letters back you are asked to go, the harder the game becomes.

The purpose of the game is to train you to be able to hold an increased number of things in your working memory at once. Jaeggi and her team wanted to know whether doing this difficult task would actually have transferable outcomes on a type of Raven test. They were astonished by their findings. After decades of research ruling out transferable gains from tests intending to improve cognitive capacity, they had done it. They found that before doing the *n*-back, on average participants could solve 9 or 10 of the 29 Raven questions in 10 minutes, but this was increased by 4.4 questions after 19 days of *n*-back training. Participants could now keep more items in their working memory, as demonstrated by their enhanced performance on the Raven intelligence test.

This study opened up researchers, who had for years largely failed to find transferable gains, to the idea that perhaps we *could* make brain boosting games that actually improve people's lives. But the idea came crashing down once again when these fluid intelligence boosting effects seemed few and far between and other researchers were not able to find the same results. In 2015 Monica Melby-Lervåg from the University of Oslo, who studies how we can help people with special needs, and her colleague Charles Hulme, a psychological scientist,[4] conducted a meta-analysis of whether working memory training is effective. They looked at all the research that had ever been conducted on the topic, and through a statistical analysis of all the results put together, they came to the conclusion that while our trusty iPhone games are likely to make us better gamers, to date 'there is no convincing evidence that working memory training produces general cognitive benefits'.[5]

While memory-enhancing games are an exciting avenue to explore, it seems premature for us to go home and tirelessly play games to make our memories better.

Secret agent mnemonics

Occasionally organisations approach me as a consultant, with a view to helping their members improve their memories. Perhaps the most interesting work of this kind that I have been involved with has been with military institutions.

When you think about how a memory specialist might be able to help the military you may be tempted to think of thriller-like scenarios, perhaps using memory research to improve the interrogation of terrorists, or to change the way the military questions civilian witnesses. You may even jump to my favourite question that people sometimes ask me on this subject, 'Do you implant false memories in spies?' I'm sorry to disappoint you but no, I do not. While I'm sure that some of my colleagues have been directly involved in helping the military understand the psychology of interrogation, and even in spy training, I have been approached mostly about one thing: teaching operatives how to identify and retain high-quality information.

My role has not been to help the military squeeze information out of others, but to help them use evidence-based techniques for their own remembering. The military has strong awareness that they need their operatives to come home with reliable memories. Talking to informants embedded in unfamiliar cultures, military agents get little, if any, time to take notes. They need to know how memory works and how they can best avoid memory illusions, as the stakes are incredibly high and to make good decisions they need to have intelligence they can trust. In order to prevent them from making basic mistakes with their own memories I teach them many of the things discussed in this book: how biases can taint the perception of information; how memories are adaptive to social demands; and how memory naturally changes over time.

They need to use memory tools to help them remember on the ground – after all, the kind of information they are collecting is hard to come by and they need to avoid it becoming in any way corrupted. So, how do I help them with this? I teach them a few simple mnemonics. Mnemonics are any kind of simple technique to make it easier for us to remember a specific piece of information. They can take many forms – rhymes, acronyms or mental imagery, to name but a few.

According to a memory mnemonic master, Ed Cooke, 'What you have to understand is that even average memories are remarkably powerful if used properly.'[6] Cooke has become a memory 'Grandmaster', for which you need to demonstrate to the World Memory Sport Council (yes, it exists) that you are able to do three things: memorise 1,000 random digits in an hour, memorise the order of ten decks of cards in an hour, and memorise the order of a deck of cards in under two minutes. It is a feat that seems impossible to us mere mortals. Luckily for us, Cooke has dedicated his life to understanding and applying amazing memory techniques. He performs these feats not because of some innate ability but through self-taught mnemonics. Mnemonics that you too can master.

Mnemonics have been around probably as long as

self-reflective memory itself. You probably learned quite a few yourself as a child, perhaps to help you remember the dates of certain historical events, or perhaps the names of planets in the solar system. One of the mnemonics I learned in primary school is 'Never Eat Soggy Wieners'. It was a mnemonic device for me to remember the four cardinal directions: North East South West. It's silly, it's juvenile – but I will never forget it. Now, even though I know the directions, I can't get rid of it. I have the opposite problem to difficulty with memory retention – I have a memory *suppression* problem, whereby I can't think about the compass points *without* thinking of that phrase. But as a memory aid it worked like gold.

It worked so well because the sentence is conceptually easy and makes grammatical sense, but more importantly it's also quite weird and you'd be highly unlikely to hear it in day-to-day life.

Be weird

Research clearly shows that, from a memory perspective, weirdness sticks, or to put it differently, unexpected components generally make for the most memorable examples and pieces of information.

Consider a statement like: 'Don't think about pink elephants.' It is something you can visualise. It is unexpected in a conversation. It is a bit weird. We probably have few previous associations, if any, with pink elephants. This particular statement has the added appeal of automatically making you do exactly the opposite of what it is telling you. For the rest of this chapter, perhaps even long after reading it, you will probably remember that it talked about pink elephants. We don't care about the pink elephants themselves; we are using them as a mnemonic device to remember the effectiveness of unusual examples for helping us remember things. See what I did there? I just used a mnemonic device to help you remember something about mnemonic devices.

According to research from 2013 published by psychological scientist Lisa Geraci from Texas A&M University and her colleagues,[7] the bizarreness effect – our tendency to have a better memory for the unusual – is well documented. As they put it, 'this bizarreness effect is a robust finding in recall that has been obtained across a variety of encoding tasks and delays.'

In the typical research paradigm looking into this effect, partici pants study sentences, some of which are bizarre and some of which are more commonplace. Within these sentences are embedded nouns written in capital letters. The participants are subsequently asked to report back all the nouns they can remember from the sentences in a free recall task. It turns out that nouns embedded within a weird sentence like 'The DOG rode the BICYCLE down the STREET' are recalled significantly better than the same nouns given in a typical sentence, such as 'The DOG chased the BICYCLE down the STREET.' Similarly, a sentence like 'The BISCUITS screamed when the OVEN jumped out the WINDOW' will force you to put more effort into making a connection between the words, and to visualise them more, than 'The BISCUITS were visible through the OVEN WINDOW'.

Geraci and her team conducted exactly this kind of experiment. Normally in this kind of study participants are given lists in which half the sentences are bizarre and half are normal. But Geraci had participants work either with a mix of both normal and bizarre sentences or with sets of sentences that were either all normal or all bizarre. What she found was that the memory benefits associated with bizarre sentences only emerged when the participants were presented with a mix of sentences, so the bizarreness of a sentence is no longer useful when everything we need to remember is bizarre. It seems common sense when you think about it – part of what makes the weirdness work is that it stands in contrast to the normality of everything else.

But why does weirdness work at all? It has to do with our memory being associative by nature. If we think back to the conception of memory as a giant web of connected memory fragments in the brain, what mnemonics help us to do is make more associations than we otherwise would.

For example, let's imagine we need to remember COUCH, BLUE and KEVIN. We are most likely to remember this if we involve as many faculties as possible to make a good network of memory associations. In line with our associative memory networks, discussed in Chapter 3, trying to make a multisensory memory both increases the amount of effort we are putting into making it stick, and makes us create as far-reaching a network as possible. So, to make our memory we could imagine feeling the couch cushions, picture the colour as a vivid blue, and imagine hearing our friend Kevin yelling at us. If we want to add bizarreness to help it stick, we may picture Kevin with a supersized neon blue couch.

By creating this kind of multisensory picture we are engaging many more parts of our brain than if we were to try to remember the words alone. Instead of just using the parts responsible for language, we are now actively making connections in the regions of our brain responsible for vision, touch and hearing. If we want to further improve this mnemonic, according to a review of the effect of imagery on memory recall in 2012 by cognitive psychologist Kristin Morrison and her team at Georgia Institute of Technology,[8] 'the more *interacting* and vivid a bizarre picture is, the more likely it will be recalled'. This means that we want our Kevin to be angrily painting his supersized neon blue couch. Interactivity matters, as it creates even more associations between concepts.

The more potential pathways there are to a memory (the more associative links), the faster and more likely we are to get to where we need to when we try to recall it. This is really the core principle behind most memory aids; we want to make things vivid, bizarre, and part of a situation. That's why the Never Eat Soggy Wieners acrostic from my childhood worked, it was weird and easy to remember and I could build on it to trigger my memory of North East South West.

Another popular mnemonic is the 'memory palace'. A memory palace, also known as the method of loci, is when we use a place for which we already have memories to build associations on. Typically this method involves forming a strong memory of a house, our palace, where we picture the exact layout and how each room is decorated. We then use that real memory as a place to store our memories. It's a bit like having a virtual memory world where we can store real memories. When we picture ourselves going through that virtual world, we can then leave things in it, along the lines of 'I'll just put this memory here for now.'

For example, we may need to remember to get eggs, yellow paint, and three spatulas from the store. To do this we might walk through our memory palace and 'leave' the eggs by the door on the mat, walk into the house and find the wall to our left dripping in paint, and then stumble over three spatulas attacking us as we go into the living room. When we then want to remember these items we just need to walk through our house again to find the items where we left them.

According to author and mnemonics expert Joshua Foer, in order to remember our objects in the locations we ideally want to create a 'comically surreal, and unforgettable, memory palace'. This works because it capitalises on the bizarreness effect, and it maximises the associations between concepts we already have in memory (our palace) with new things that we want to remember (our shopping list).

You can see how these mnemonics, just like others that you will come across in books that propose to improve your memory, are all using associations and weirdness to their advantage. And now that you know, so can you. Go forth and be weird. Your memory will thank you.

I prefer my version of the truth

'How do you live with it?' This is a question I get asked all the time. It seems to be asking how I can refrain from being in a state of constant despair, knowing that I cannot trust my memory. As one of my undergraduate students once said when I first began talking about this topic 'I don't even know what is *real* anymore!'

Can we be happy knowing that our memories are highly questionable? Absolutely. Happier, I would argue. We are now less likely to be a victim of our own memories, and can assume at least some control over this elusive process. It can be distressing to think that all our memories are tainted in minor or even major ways. But this introduces a flexible creativity into reality. Memory is personal and subjective anyway, so when we are in the surprisingly common situation of being faced with multiple interpretations or versions of what happened and have no independent evidence to help us know what actually happened, we can pick the one we like best. We all prefer our versions of the truth, but when we understand memory processes we can actively weave the life we want in ways that maximise our happiness and the happiness of those around us. It allows us to treat life with a sense of magical realism, as a 'paint by numbers' of reality.

Understanding the fallibility of memory also allows us to avoid the marketing policies that try to take advantage of our innate biases, like the 'subscribe now, pay later' models explored in an earlier chapter. It makes us less likely to be overconfident in our memory abilities. This allows us to be on our toes, so we can make decisions that are actually far more beneficial to ourselves, and less likely to be tainted by biases that we typically don't even know are being activated. Being critical of our memories thus makes us better consumers of information and of actual goods.

It also lets us better understand our everyday disputes with friends and family or, if they become public, with the media – à la Brian Williams. People we once may have assumed were lying are now critically approached with compassion. We know that people can get memories fundamentally wrong, believing things that never actually happened. And precisely when versions of memories fit well with who we think we are, or who we want to be, these memories may be extra likely to become part of our perceived personal past.

Remember the story of Brian Williams, the disgraced newscaster we talked about in Chapter 7? He probably liked the idea that he was attacked by a WMD while in a helicopter, so when he began to misremember he was probably less critical of this memory than he could have been. While we can never be sure whether someone is lying to us or not, we at least now acknowledge that such situations could result from an act of misremembering.

This knowledge also informs us about how those involved with the legal system, including victims, witnesses, suspects, and even the police themselves, can get their memories muddled. It makes us critical of accepting accounts as reliable and true when they are not corroborated with independent evidence. As the world's single most influential false memory expert Elizabeth Loftus said in her fantastic TED talk in 2013,[9] 'Most people cherish their memories, know that they represent their identity, who they are, where they came from. And I appreciate that. I feel that way too. But I know from my work how much fiction is already in there. If I've learned anything from these decades of working on these problems, it's this: just because somebody tells you something and they say it with confidence, just because they say it with lots of detail, just because they express emotion when they say it, it doesn't mean that it really happened.' This knowledge has the power to revolutionise the legal system and to help prevent miscarriages of justice.

Knowing that our memories are unreliable also inspires us to seek out exactly how and when memory processes break down. For me it has proven an insanely fascinating ride, trying to figure out these memory illusions in the lab and to generate practical applications for the police, the military and the business world. I hope that for you this also proves true, as you can search for applications far and wide and open a Pandora's box of potential fascination and intrigue, allowing you to see as amazing a process we too often take for granted. How and why we remember is a topic that never gets old.

Finally, understanding all the shortcomings that our memory system presents allows us to adhere to a whole new ethos. Our past is a fictional representation, and the only thing we can be even somewhat sure of is what is happening now. It encourages us to live in the moment and not to place too much importance on our past. It forces us to accept that the best time of our lives, and our memory, is right now.

And so I leave you. I hope you take all that you have learned from this book onwards and forwards. Spread the word about memory illusions and utilise your new-found insight into our memory processes to make your everyday life just a little bit better.

Acknowledgements

Fred. Without you I would not be the academic I am today. I would have dropped out in first year, run away to study fine art, and would have never written this book.

Family made this book possible.
Mom. Without you I would not be the woman I am today.
Mark. Without you I would not be the artist I am today.
Omi. Without you I would not be the lady I am today.
Dad. Without you I would not be the intellectual I am today.

Friends made this book possible.
Noemi Drecksler, John Gaspar, Annelies Vredeveldt, Mara Toebbens, Sophie van der Zee, Jodie Perzan, Bianca Baker. Your support kept me going.

Academic mentors made this book possible.
Steve Hart. Without you I would have never studied forensic psychology.
Steve Porter. Without you I would have never researched false memories.
Ray Bull. Without you I would have never been so confident in my intelligence.
Barry Beyerstein. Without you I would have never become a skeptic.
Elizabeth Loftus. Without you there would be no applied false memory science.

Those directly involved made this book possible.

Kirsty McLachlan at DGA.
Your fortuitous reading of the *Evening Standard*, and blind faith in me, made this book possible.

Harry Scoble at Random House.
Your belief in me as a writer, coupled with many months of editing, made this book possible.

The team at DGA.
Your belief in this book, and perseverance, sold this book ten times before it was even written.

Christian Koth at Hanser.
Your instant purchase of the book created a crazier storm of offers than I could have ever imagined.

Endnotes

Chapter 1

1 http://www.theguardian.com/notesandqueries/query/0,,-2899,00.html

2 Nahum, L., A. Bouzerda-Wahlen, A., A. Guggisberg, A., Ptak, R., & Schnider, A. (2012). Forms of confabulation: dissociations and associations. *Neuropsychologia*, 50(10): 2524–34.

3 Hyman Jr, I. E., & Pentland, J. (1996). The role of mental imagery in the creation of false childhood memories. *Journal of Memory and Language*, 35(2): 101–17.

4 Miller, G. A. (1956). The magical number seven, plus or minus two: some limits on our capacity for processing information. *Psychological Review*, 63(2): 81.

5 Cowan, N. (2001). The magical number 4 in short-term memory: a reconsideration of mental storage capacity. *Behavioral and Brain Sciences*, 24: 87–185.

6 Tamnes, C. K., Walhovd, K. B., Grydeland, H., Holland, D., Østby, Y., Dale, A. M., & Fjell, A. M. (2013). Longitudinal working memory development is related to structural maturation of frontal and parietal cortices. *Journal of Cognitive Neuroscience*, 25(10): 1611–23.

7 http://www.nobelprize.org/nobel_prizes/medicine/laureates/1949/moniz-article.html

8 Freeman, W. (1967). Multiple lobotomies. *American Journal of Psychiatry*, 123(11): 1450–2.

9 Miller, G. A. (1956). The magical number seven, plus or minus two: some limits on our capacity for processing information. *Psychological Review*, 63(2): 81.

10 Wang, Q., & Peterson, C. (2014). Your earliest memory may be earlier than you think: prospective studies of children's dating of earliest childhood memories. *Developmental Psychology*, 50(6): 1680.

11 Miles, C. (1895). A study of individual psychology. *American Journal of Psychology*, 6: 534–58.

12 http://news.harvard.edu/gazette/2002/11.07/01-memory.html

13 When Do Babies Develop Memories? http://abcnews.go.com/Technology/story?id=97848

14 Lie, E., & Newcombe, N. S. (1999). Elementary school children's explicit and implicit memory for faces of preschool classmates. *Developmental Psychology*, 35(1): 102.

15 Knickmeyer, R. C., Gouttard, S., Kang, C., Evans, D., Wilber, K., Smith, J. K., et al. (2008). A structural MRI study of human brain development from birth to 2 years. *Journal of Neuroscience*, 28(47): 12176–82.

16 Caviness Jr, V. S., Kennedy, D. N., Richelme, C., Rademacher, J., & Filipek, P. A. (1996). The human brain age 7–11 years: a volumetric analysis based on magnetic resonance images. *Cerebral Cortex*, 6(5): 726–36.

17 Abitz, M., Nielsen, R. D., Jones, E. G., Laursen, H., Graem, N., & Pakkenberg, B. (2007). Excess of neurons in the human newborn mediodorsal thalamus compared with that of the adult. *Cerebral Cortex*, 17(11): 2573–8.

18 Huttenlocher, P. R. (1990). Morphometric study of human cerebral cortex development. *Neuropsychologia*, 28(6): 517–27.

19 Chechik, G., Meilijson, I., & Ruppin, E. (1998). Synaptic pruning in development: a computational account. *Neural Computation*, 10(7): 1759–77.

20 Erdelyi, M. H. (1994). In Memoriam to Dr. Nicholas P. Spanos. *International Journal of Clinical and Experimental Hypnosis*, 42(4).

21 Spanos, N. P., Burgess, C. A., Burgess, M. F., Samuels, C., & Blois, W. O. (1999). Creating false memories of infancy with hypnotic and non-hypnotic procedures. *Applied Cognitive Psychology*, 13(3): 201–18.

22 Spanos, N. P., Burgess, C. A., & Burgess, M. F. (1994). Past-life identities, UFO abductions, and satanic ritual abuse: the social construction of memories. *International Journal of Clinical and Experimental Hypnosis*, 42(4): 433–46.

23 Braun, K. A., Ellis, R., & Loftus, E. F. (2002). Make my memory: How advertising can change our memories of the past. *Psychology & Marketing*, 19(1): 1–23.

24 Strange, D., Sutherland, R., & Garry, M. (2006). Event plausibility does not determine children's false memories. *Memory*, 14(8): 937–51.

25 For more information visit arhopwood.com and falsememoryarchive.com.

26 Flavell, J. H., & Wellman, H. M. (1975). Metamemory.

27 Shonkoff, J. P., Garner, A. S., Siegel, B. S., Dobbins, M. I., Earls, M. F., McGuinn, L., et al. (2012). The lifelong effects of early childhood adversity and toxic stress. *Pediatrics*, 129(1): e232–e246.

Chapter 2

1 http://www.independent.co.uk/news/science/remember-the-dress-brain-scientists-now-see-the-internet-meme-as-an-invaluable-research-tool-10251422.html

2 Lafer-Sousa, R., Hermann, K. L., & Conway, B. R. (2015). Striking individual differences in color perception uncovered by 'the dress' photograph. *Current Biology*, 25(13): R545–R546.

3 Gegenfurtner, K. R., Bloj, M., & Toscani, M. (2015). The many colours of 'the dress'. *Current Biology*, 25(13): R543–R544.

4 Winkler, A. D., Spillmann, L., Werner, J. S., & Webster, M. A. (2015). Asymmetries in blue–yellow color perception and in the color of 'the dress'. *Current Biology*, 25(13): R547–R548.

5 Gibson, J. J., & Gibson, E. J. (1955). Perceptual learning: differentiation or enrichment? *Psychological Review*, 62(1): 32–41.

6 Korva, N., Porter, S., O'Connor, B. P., Shaw, J., & ten Brinke, L. (2013). Dangerous decisions: Influence of juror attitudes and defendant appearance on legal decision-making. *Psychiatry, Psychology and Law*, 20(3): 384–98.

7 Ambady, N., & Rosenthal, R. (1992). Thin slices of expressive behavior as predictors of interpersonal consequences: a meta-analysis. *Psychological Bulletin*, 111(2): 256–74.

8 Cahill, L., & McGaugh, J. L. (1998). Mechanisms of emotional arousal and lasting declarative memory. *Trends in Neurosciences*, 21(7): 294–9.

9 Cahill, L., & McGaugh, J. L. (1995). A novel demonstration of enhanced memory associated with emotional arousal. *Consciousness and Cognition*, 4(4): 410–21.

10 Schilling, T. M., et al. (2013). For whom the bell (curve) tolls: cortisol rapidly affects memory retrieval by an inverted U-shaped dose–response relationship. *Psychoneuroendocrinology* 38(9): 1565–72.

11 http://www.apa.org/science/about/psa/2012/02/emotional-arousal.aspx

12 Pearce, S. A., Isherwood, S., Hrouda, D., Richardson, P. H., Erskine, A.,

& Skinner, J. (1990). Memory and pain: tests of mood congruity and state dependent learning in experimentally induced and clinical pain. *Pain*, 43(2): 187–93.

13 Tulving, E. (2002). Chronesthesia: Conscious awareness of subjective time.

14 Kahneman, D., & Tversky, A. (1977). Intuitive prediction: Biases and corrective procedures. DECISIONS AND DESIGNS INC MCLEAN VA.

15 Buehler, R., Griffin, D., & Peetz, J. (2010). The planning fallacy: cognitive, motivational, and social origins. *Advances in Experimental Social Psychology*, 43, 1–62.

16 Also important: Buehler, R., Griffin, D., & Ross, M. (1994). Exploring the 'planning fallacy': why people underestimate their task completion times. *Journal of Personality and Social Psychology*, 67(3): 366.

17 Tobin, S., Bisson, N., & Grondin, S. (2010). An ecological approach to prospective and retrospective timing of long durations: a study involving gamers. *PloS One*, 5(2): e9271.

18 Gaskell, G. D., Wright, D. B., & O'Muircheartaigh, C. A. (2000). Telescoping of landmark events: implications for survey research. *Public Opinion Quarterly*, 64(1): 77–89.

19 Janssen, S., Chessa, A., & Murre, J. (2005). The reminiscence bump in autobiographical memory: effects of age, gender, education, and culture. *Memory*, 13(6): 658–68.

20 Conway, M. A., Wang, Q., Hanyu, K., & Haque, S. (2005). A cross-cultural investigation of autobiographical memory: on the universality and cultural variation of the reminiscence bump. *Journal of Cross-Cultural Psychology*, 36(6): 739–49.

Chapter 3

1 Hunt, K. L., & Chittka, L. (2015). Merging of long-term memories in an insect. *Current Biology*, 25(6): 741–5.

2 Hunt, K., & Chittka, L. (2014). False memory susceptibility is correlated with categorisation ability in humans. *F1000Research*, 3.

3 Baudry, M., Bi, X., Gall, C., & Lynch, G. (2011). The biochemistry of memory: The 26 year journey of a 'new and specific hypothesis'. *Neurobiology of learning and memory*, 95(2), 125-133.

4 http://www.genomenewsnetwork.org/articles/2004/01/09/memories.php

5 Fioriti, L., Myers, C., Huang, Y. Y., Li, X., Stephan, J. S., Kandel, E. R., et al. (2015). The persistence of hippocampal-based memory requires protein synthesis mediated by the prion-like protein CPEB3. *Neuron*, 86(6): 1433–48.

6 Stephan, J. S., Fioriti, L., Lamba, N., Colnaghi, L., Karl, K., Derkatch, I. L., & Kandel, E. R. (2015). The CPEB3 protein is a functional prion that interacts with the actin cytoskeleton. *Cell Reports*, 11(11): 1772–85.

7 http://www.scientificamerican.com/article/prions-are-key-to-preserving-long-term-memories/

8 Nader, K., Schafe, G. E., & Le Doux, J. E. (2000). Fear memories require protein synthesis in the amygdala for reconsolidation after retrieval. *Nature*, 406(6797): 722–6.

9 Chan, J. C., & LaPaglia, J. A. (2013). Impairing existing declarative memory in humans by disrupting reconsolidation. *Proceedings of the National Academy of Sciences*, 110(23): 9309–13.

10 Beracochea, D. (2006). Anterograde and retrograde effects of benzodiazepines on memory. *Scientific World Journal*, 6, 1460–5.

11 Vidailhet, P., Danion, J. M., Kauffmann-Muller, F., Grangé, D., Giersch, A., Van Der Linden, M., & Imbs, J. L. (1994). Lorazepam and diazepam effects on memory acquisition in priming tasks. *Psychopharmacology*, 115(3): 397–406.

12 Vidailhet, P., Danion, J. M., Chemin, C., & Kazès, M. (1999). Lorazepam impairs both visual and auditory perceptual priming. *Psychopharmacology*, 147(3): 266–73.

13 de Lavilléon, G., Lacroix, M. M., Rondi-Reig, L., & Benchenane, K. (2015). Explicit memory creation during sleep demonstrates a causal role of place cells in navigation. *Nature Neuroscience*.

14 O'Keefe, J., & Dostrovsky, J. (1971). The hippocampus as a spatial map. Preliminary evidence from unit activity in the freely-moving rat. *Brain Research*, 34(1): 171–5.

15 Ramirez, S., Liu, X., Lin, P. A., Suh, J., Pignatelli, M., Redondo, R. L., et al. (2013). Creating a false memory in the hippocampus. *Science*, 341(6144): 387–91.

16 http://www.theguardian.com/education/2015/sep/16/what-happens-in-your-brain-when-you-make-a-memory

17 Ibsen, S., Tong, A., Schutt, C., Esener, S., & Chalasani, S. H. (2015). Sonogenetics is a non-invasive approach to activating neurons in *Caenorhabditis elegans*. *Nature Communications*, 6.

18 Ebbinghaus, H. (1885). *Über das Gedächtnis* [On memory]. Leipzig, Germany: Duncker and Humblot.

19 Ebbinghaus, H. (1913). *Memory: A Contribution to Experimental Psychology* (No. 3). University Microfilms.

20 Brainerd, C. J., & Reyna, V. F. (2002). Fuzzy-trace theory and false memory. *Current Directions in Psychological Science*, 11(5): 164–9.

Chapter 4

1 Parker, E. S., Cahill, L., & McGaugh, J. L. (2006). A case of unusual autobiographical remembering. *Neurocase*, 12(1): 35–49.

2 Ericsson, K. A., Delaney, P. F., Weaver, G., & Mahadevan, R. (2004). Uncovering the structure of a memorist's superior 'basic' memory capacity. *Cognitive Psychology*, 49(3): 191–237.

3 Patihis, L., Frenda, S. J., LePort, A. K., Petersen, N., Nichols, R. M., Stark, C. E., et al. (2013). False memories in highly superior autobiographical memory individuals. *Proceedings of the National Academy of Sciences*, 110(52): 20947–52.

4 http://nymag.com/scienceofus/2014/11/what-its-like-to-remember-almost-everything.html#

5 Penfield, W. (1952). Memory mechanisms. *AMA Archives of Neurology & Psychiatry*, 67(2): 178–98.

6 Penfield, W., & Perot, P. (1963). The brain's record of auditory and visual experience. *Brain*, 86(4): 595–696.

7 Milner, B. (1977). Wilder Penfield: his legacy to neurology. Memory mechanisms. *Canadian Medical Association Journal*, 116(12): 1374.

8 http://www.scientificamerican.com/article/is-there-such-a-thing-as/

9 Searleman, A., & Herrmann, D. J. (1994). *Memory from a Broader Perspective.* New York: McGraw-Hill.

10 Gray, C. R., & Gummerman, K. (1975). The enigmatic eidetic image: a critical examination of methods, data, and theories. *Psychological Bulletin*, 82(3): 383–407.

11 Haber, R. N. (1979). Twenty years of haunting eidetic imagery: where's the ghost?. *Behavioral and Brain Sciences*, 2(04): 583–94.

12 Giray, E. F., Altkin, W. M., Roodin, P. A., & Vaught, G. M. (1977). The enigmatic eidetic image: a reply to Gray and Gummerman. *Perceptual and Motor Skills*, 44(1): 191–4.

13 Patihis, L., Frenda, S. J., LePort, A. K., Petersen, N., Nichols, R. M., Stark, C. E., et al. (2013). False memories in highly superior autobiographical memory individuals. *Proceedings of the National Academy of Sciences*, 110(52): 20947–52.

14 Roediger, H. L., & McDermott, K. B. (1995). Creating false memories: Remembering words not presented in lists. *Journal of Experimental Psychology: Learning, Memory, and Cognition*, 21(4): 803.

15 Collins, A. M., & Loftus, E. F. (1975). A spreading-activation theory of semantic processing. *Psychological Review*, 82(6): 407–28.

16 http://www.nytimes.com/2011/10/16/books/review/is-the-brain-good-at-what-it-does.html?_r=0

17 http://singularityhub.com/2011/09/29/hyperthymesia-%E2%80%93-a-newly-discovered-memory-in-which-people-remember-every-day-of-their-lives-video/

18 Treffert, D. A. (2009). The savant syndrome: an extraordinary condition. A synopsis: past, present, future. *Philosophical Transactions of the Royal Society of London B: Biological Sciences*, 364(1522): 1351–7.

19 Peek, F., & Hanson, L. (2007). *The Life and Message of the Real Rain Man: The Journey of a Mega-savant*. National Professional Resources Inc./Dude Publishing.

20 Bauman, M., & Kemper, T. L. (1985). Histoanatomic observations of the brain in early infantile autism. *Neurology*, 35(6): 866–74.

21 Maier, S., van Elst, L. T., Beier, D., Ebert, D., Fangmeier, T., Radtke, M., et al. (2015). Increased hippocampal volumes in adults with high functioning autism spectrum disorder and an IQ>100: A manual morphometric study. *Psychiatry Research: Neuroimaging*, 234(1): 152–5.

22 Shalom, D. B. (2009). The medial prefrontal cortex and integration in autism. *Neuroscientist*, 15(6): 589–98.

23 Baron-Cohen, S. (1997). *Mindblindness: An Essay on Autism and Theory of Mind*. MIT Press.

24 http://www.npr.org/sections/health-shots/2013/12/18/255285479/when-memories-never-fade-the-past-can-poison-the-present

25 http://nymag.com/scienceofus/2014/11/what-its-like-to-remember-almost-everything.html

26 http://gizmodo.com/how-memory-hacking-is-becoming-a-reality-1757888568

27 Kuhl, B. A., Dudukovic, N. M., Kahn, I., & Wagner, A. D. (2007). Decreased demands on cognitive control reveal the neural processing benefits of forgetting. *Nature Neuroscience*, 10(7): 908–14.

28 Cottencin, O., Vaiva, G., Huron, C., Devos, P., Ducrocq, F., Jouvent, R., et al. (2006). Directed forgetting in PTSD: a comparative study versus normal controls. *Journal of Psychiatric Research*, 40(1): 70–80.

Chapter 5

1 Arzi, A., Shedlesky, L., Ben-Shaul, M., Nasser, K., Oksenberg, A., Hairston, I. S., & Sobel, N. (2012). Humans can learn new information during sleep. *Nature Neuroscience*, 15(10): 1460–5.

2 Rauscher, F. H., Shaw, G. L., & Ky, K. N. (1993). Music and spatial task performance. *Nature*, 365(6447): 611.

3 DeLoache, J. S., Chiong, C., Sherman, K., Islam, N., Vanderborght, M., Troseth, G. L., et al. (2010). Do babies learn from baby media? *Psychological Science.*

4 Zimmerman, F. J., Christakis, D. A., & Meltzoff, A. N. (2007). Associations between media viewing and language development in children under age 2 years. *Journal of Pediatrics*, 151(4): 364–8.

5 https://www.aap.org/en-us/advocacy-and-policy/aap-health-initiatives/pages/media-and-children.aspx

6 Simons, D. J., & Chabris, C. F. (1999). Gorillas in our midst: sustained inattentional blindness for dynamic events. *Perception*, 28(9): 1059–74.

7 Simons, D. J., & Levin, D. T. (1998). Failure to detect changes to people during a real-world interaction. *Psychonomic Bulletin & Review*, 5(4): 644–9.

8 Hyman, I. E., Boss, S. M., Wise, B. M., McKenzie, K. E., & Caggiano, J. M. (2010). Did you see the unicycling clown? Inattentional blindness while walking and talking on a cell phone. *Applied Cognitive Psychology*, 24(5): 597–607.

9 https://www.psychologytoday.com/blog/mental-mishaps/201004/failing-notice-haircuts-missing-buildings-and-changed-conversation

10 Levin, D. T., Momen, N., Drivdahl IV, S. B., & Simons, D. J. (2000). Change blindness blindness: The metacognitive error of overestimating change-detection ability. *Visual Cognition*, 7(1–3): 397–412.

11 Feld, G. B., & Diekelmann, S. (2015). Sleep smart – optimizing sleep for declarative learning and memory. *Frontiers in Psychology*, 6: 622.

12 Wang, G., Grone, B., Colas, D., Appelbaum, L., & Mourrain, P. (2011). Synaptic plasticity in sleep: learning, homeostasis and disease. *Trends in Neurosciences*, 34(9): 452–63.

13 Yang, G., Lai, C. S. W., Cichon, J., Ma, L., Li, W., & Gan, W. B. (2014). Sleep promotes branch-specific formation of dendritic spines after learning. *Science*, 344(6188): 1173–8.

14 http://www.newyorker.com/magazine/1933/10/07/talk-in-dreams

15 https://patentimages.storage.googleapis.com/pages/US1886358-0.png

16 http://www.phonophan.com/articles.html

17 https://patentimages.storage.googleapis.com/pages/US1886358-0.png

18 Sucala, M., Schnur, J. B., Glazier, K., Miller, S. J., Green, J. P., & Montgomery, G. H. (2013). Hypnosis – there's an app for that: a systematic review of hypnosis apps. *International Journal of Clinical and Experimental Hypnosis*, 61(4): 463–74.

19 Simon, C. W., & Emmons, W. H. (1956). EEG, consciousness, and sleep. *Science*, 124(3231): 1066–9.

20 Hennevin, E., Hars, B., Maho, C., & Bloch, V. (1995). Processing of learned information in paradoxical sleep: relevance for memory. *Behavioural Brain Research*, 69(1): 125–35.

21 Cordi, M. J., Diekelmann, S., Born, J., & Rasch, B. (2014). No effect of odor-induced memory reactivation during REM sleep on declarative memory stability. *Frontiers in Systems Neuroscience*, 8.

22 Mazzoni, G., Laurence, J. R., & Heap, M. (2014). Hypnosis and memory: Two hundred years of adventures and still going! *Psychology of Consciousness: Theory, Research, and Practice*, 1(2): 153.

23 Sheehan, P. W. & Perry, C. W. (2015). *Methodologies of Hypnosis* (Psychology Revivals)*: A Critical Appraisal of Contemporary Paradigms of Hypnosis.* Routledge.

24 Montgomery, G. H., David, D., Winkel, G., Silverstein, J. H., & Bovbjerg, D. H. (2002). The effectiveness of adjunctive hypnosis with surgical patients: a meta-analysis. *Anesthesia & Analgesia*, 94(6): 1639–45.

25 Gonsalkorale, W. M., Houghton, L. A., & Whorwell, P. J. (2002). Hypnotherapy in irritable bowel syndrome: a large-scale audit of a clinical service with examination of factors influencing responsiveness. *American Journal of Gastroenterology*, 97(4): 954–61.

26 Castel, A., Salvat, M., Sala, J., & Rull, M. (2009). Cognitive-behavioural

group treatment with hypnosis: a randomized pilot trail in fibromyalgia. *Contemporary Hypnosis*, 26(1): 48–59.

27 http://psychcentral.com/news/2012/10/06/not-getting-sleepy-not-everyone-can-be-hypnotized/45672.html

28 Wagstaff, G. F. (1997). What is hypnosis? *Interdisciplinary Science Reviews*, 22(2): 155–63.

29 Barber, T. X. (1962). Hypnotic age regression: a critical review. *Psychosomatic Medicine*, 24(3): 286–99.

30 Sargant, W. (1957). *Battle for the Mind: A Physiology of Conversion and Brain-washing.* London: Heinemann.

31 Rahnev, D. A., Huang, E., & Lau, H. (2012). Subliminal stimuli in the near absence of attention influence top-down cognitive control. *Attention, Perception & Psychophysics*, 74(3): 521–32.

32 Vokey, J. R., & Read, J. D. (1985). Subliminal messages: Between the devil and the media. *American Psychologist*, 40(11): 1231–9.

Chapter 6

1 Chaplin, C., & Shaw, J. (2015). Confidently Wrong: Police Endorsement of Psycho-Legal Misconceptions. *Journal of Police and Criminal Psychology*, 1-9.

2 Figures accessed on March 30TH, 2016 from http://www.innocenceproject.org

3 Svenson, O. (1981). Are we all less risky and more skillful than our fellow drivers? *Acta Psychologica*, 47(2): 143–8.

4 Johnson, D. D. P., & Fowler, J. H. (2011). The evolution of overconfidence. *Nature*, 477(7364): 317–20.

5 http://www.cozi.com/live-simply/who-works-harder-mom-or-dad

6 Amin, G. S., & Kat, H. M. (2003). Welcome to the dark side: hedge fund attrition and survivorship bias over the period 1994–2001. *Journal of Alternative Investments*, 6: 57–3.

7 Pronin, E., Kruger, J., Savtisky, K., & Ross, L. (2001). You don't know me, but I know you: the illusion of asymmetric insight. *Journal of Personality and Social Psychology*, 81(4): 639.

8 Holman, J., & Zaidi, F. (2010). The economics of prospective memory. *Available at SSRN 1662183.*

9 Kornell, N. (2011). Failing to predict future changes in memory: A stability bias yields long-term overconfidence. In A. S. Benjamin (ed.), *Successful Remembering and Successful Forgetting: A Festschrift in Honor of Robert A. Bjork.* New York, NY: Psychology Press. 365–86.

10 Koriat, A., Bjork, R. A., Sheffer, L., & Bar, S. K. (2004). Predicting one's own forgetting: the role of experience-based and theory-based processes. *Journal of Experimental Psychology: General,* 133(4): 643–56.

11 https://www.psychologytoday.com/blog/everybody-is-stupid-except-you/201008/long-term-overconfidence

12 Furl, N., Garrido, L., Dolan, R. J., Driver, J., & Duchaine, B. (2011). Fusiform gyrus face selectivity relates to individual differences in facial recognition ability. *Journal of Cognitive Neuroscience,* 23(7): 1723–40.

13 Sacks, O. (1998). *The Man Who Mistook His Wife for a Hat and Other Clinical Tales.* Simon and Schuster.

14 Kennerknecht, I., Grueter, T., Welling, B., Wentzek, S., Horst, J., Edwards, S., et al. (2006). First report of prevalence of non-syndromic hereditary prosopagnosia (HPA). *American Journal of Medical Genetics Part A,* 140A(15): 1617–22.

15 Russell, R., Duchaine, B., & Nakayama, K. (2009). Super-recognizers: People with extraordinary face recognition ability. *Psychonomic Bulletin & Review,* 16(2): 252–7.

16 https://www.newscientist.com/article/mg22830484-800-super-recognisers-could-be-used-to-identify-strangers-in-cctv/

17 White, D., Kemp, R. I., Jenkins, R., Matheson, M., & Burton, A. M. (2014). Passport officers' errors in face matching. *PloS One,* 9(8): e103510.

18 Palmer, M. A., Brewer, N., Weber, N., & Nagesh, A. (2013). The confidence-accuracy relationship for eyewitness identification decisions: Effects of exposure duration, retention interval, and divided attention. *Journal of Experimental Psychology: Applied,* 19(1): 55–71.

19 Blais, C., Jack, R. E., Scheepers, C., Fiset, D., & Caldara, R. (2008). Culture shapes how we look at faces. *PLoS One,* 3(8): e3022.

20 Ross, D. A., Deroche, M., & Palmeri, T. J. (2014). Not just the norm: exemplar-based models also predict face aftereffects. *Psychonomic Bulletin & Review,* 21(1): 47–70.

21 Young, S. G., Hugenberg, K., Bernstein, M. J., & Sacco, D. F. (2012). Perception and motivation in face recognition: a critical review of theories of the Cross-Race Effect. *Personality and Social Psychology Review,* 16(2): 116–42.

22 Sporer, S. L. (2001). Recognizing faces of other ethnic groups: An integration of theories. *Psychology, Public Policy, and Law*, 7(1): 36.

23 Ofshe, R., & Watters, E. (1994). *Making Monsters: False Memories, Psychotherapy, and Sexual Hysteria.* University of California Press.

Chapter 7

1 http://news.yahoo.com/nbc-news--brian-williams-recants-story-iraq-helicopter-after-soldiers-protest-231038729.html

2 Porter, S., & Birt, A. R. (2001). Is traumatic memory special? A comparison of traumatic memory characteristics with memory for other emotional life experiences. *Applied Cognitive Psychology*, 15(7): S101–S117.

3 http://www.ptsd.va.gov/professional/PTSD-overview/Dissociative_Subtype_of_PTSD.asp

4 Alpert, J. L., Brown, L. S., & Courtois, C. A. (1998). Symptomatic clients and memories of childhood abuse: What the trauma and child sexual abuse literature tells us. *Psychology, Public Policy, and Law*, 4(4): 941.

5 Staniloiu, A., & Markowitsch, H. J. (2014). Dissociative amnesia. *Lancet Psychiatry*, 1(3): 226–41.

6 Markowitsch, H. J., Kessler, J., Russ, M. O., Frölich, L., Schneider, B., & Maurer, K. (1999). Mnestic block syndrome. *Cortex*, 35(2): 219–30.

7 Porter, S., & Peace, K. A. (2007). The scars of memory: a prospective, longitudinal investigation of the consistency of traumatic and positive emotional memories in adulthood. *Psychological Science*, 18(5): 435–41.

8 Magnussen, S., & Melinder, A. (2012). What psychologists know and believe about memory: A survey of practitioners. *Applied Cognitive Psychology*, 26(1): 54–60.

9 Brown, R., & Kulik, J. (1977). Flashbulb memories. *Cognition*, 5(1): 73–99.

10 Contribution to A. R. Hopwood's False Memory Archive, 2012–14. Courtesy of the artist.

11 Day, M. V., & Ross, M. (2014). Predicting confidence in flashbulb memories. *Memory*, 22(3): 232–42.

12 Brainerd, C. J., Reyna, V. F., Wright, R., & Mojardin, A. H. (2003). Recollection rejection: false-memory editing in children and adults. *Psychological Review*, 110(4): 762.

13 Shaw, J., & Porter, S. (2015). Constructing rich false memories of committing crime. *Psychological Science*, 26(3): 291–301.

14 Hyman, I. E., Husband, T. H., & Billings, F. J. (1995). False memories of childhood experiences. *Applied Cognitive Psychology*, 9(3): 181–97.

15 Porter, S., Yuille, J. C., & Lehman, D. R. (1999). The nature of real, implanted, and fabricated memories for emotional childhood events: implications for the recovered memory debate. *Law and Human Behavior*, 23(5): 517–37.

16 Shaw, J. (2015). True or false memory? Evidence that naïve observers have difficulty identifying false memories of emotional events, especially for audio-only accounts. Paper presentation at the annual meeting of the Society for Applied Research on Memory and Cognition, Victoria, Canada.

17 Morgan, C. A., Southwick, S., Steffian, G., Hazlett, G. A., & Loftus, E. F. (2013). Misinformation can influence memory for recently experienced, highly stressful events. *International Journal of Law and Psychiatry*, 36(1): 11–17.

18 Schooler, J. W., & Engstler-Schooler, T. Y. (1990). Verbal overshadowing of visual memories: some things are better left unsaid. *Cognitive Psychology*, 22(1): 36–71.

19 Alogna, V. K., Attaya, M. K., Aucoin, P., Bahník, Š., Birch, S., Bornstein, B., et al. (2014). Registered replication report: Schooler & Engstler-Schooler (1990). *Perspectives on Psychological Science*, 9(5): 556–78.

20 Schooler & Engstler-Schooler (1990). Verbal overshadowing of visual memories.

21 Ibid.

22 Henkel, L. A. (2011). Photograph-induced memory errors: When photographs make people claim they have done things they have not. *Applied Cognitive Psychology*, 25(1): 78–86.

23 Brown, A. S., & Marsh, E. J. (2008). Evoking false beliefs about autobiographical experience. *Psychonomic Bulletin & Review*, 15(1): 186–90.

24 Wade, K. A., Garry, M., Read, J. D., & Lindsay, D. S. (2002). A picture is worth a thousand lies: using false photographs to create false childhood memories. *Psychonomic Bulletin & Review*, 9(3): 597–603.

25 Lindsay, D. S., Hagen, L., Read, J. D., Wade, K. A., & Garry, M. (2004). True photographs and false memories. *Psychological Science*, 15(3): 149–54.

26 Mitchell, J. T. (1983). When disaster strikes: the critical incident stress debriefing process. *Journal of Emergency Medical Services* 8(1): 36–9.

27 Devilly, G. J., & Cotton, P. (2003). Psychological debriefing and the workplace: Defining a concept, controversies and guidelines for intervention. *Australian Psychologist*, 38(2): 144–50.

28 Kilpatrick, D. G., Resnick, H. S., Milanak, M. E., Miller, M. W., Keyes, K. M., & Friedman, M. J. (2013). National estimates of exposure to traumatic events and PTSD prevalence using DSM-IV and DSM-5 criteria. *Journal of Traumatic Stress*, 26(5): 537–47.

Chapter 8

1 http://www.npr.org/templates/story/story.php?storyId=95256794

2 Russ, M., and Crews, D. E. (2014). A survey of multitasking behaviors in organizations. *International Journal of Human Resource Studies*, 4(1).

3 Junco, R., and Cotten S. R. (2012). No A 4 U: The relationship between multitasking and academic performance. *Computers & Education*, 59(2): 505–14.

4 Miller, E. K., & Buschman, T. J. (2013). Brain rhythms for cognition and consciousness. *Neurosciences and the Human Person: New Perspectives on Human Activities*, 121, www.casinapioiv.va/content/dam/accademia/pdf/sv121/sv121-miller.pdf

5 Buschman, T. J., Denovellis, E. L., Diogo, C., Bullock, D., & Miller, E. K. (2012). Synchronous oscillatory neural ensembles for rules in the prefrontal cortex. *Neuron*, 76(4): 838–46.

6 Strayer, D. L, Drews, F. A., & Crouch, D. J. (2006). A comparison of the cell phone driver and the drunk driver. *Human Factors*, 48(2): 381–91.

7 Miller-Ott, A. & Kelly, L. (2015). The presence of cell phones in romantic partner face-to-face interactions: An expectancy violation theory approach. *Southern Communication Journal*, 80(4): 253–70.

8 Roberts, J. A., & David, M. E. (2016). My life has become a major distraction from my cell phone: Partner phubbing and relationship satisfaction among romantic partners. *Computers in Human Behavior*, 54: 134–41.

9 Clark, B. F. (2013). From yearbooks to Facebook: public memory in transition. *International Journal of the Book*, 10(3): 19.

10 Gabbert, F., Memon, A., & Allan, K. (2003). Memory conformity: Can eyewitnesses influence each other's memories for an event? *Applied Cognitive Psychology*, 17(5): 533–43

11 Brown, A. S., Caderao, K. C., Fields, L. M. & Marsh, E. J. (2015). Borrowing personal memories. *Applied Cognitive Psychology*, 29(3): 471–7.

12 Asch, S. E. (1956). Studies of independence and conformity: A minority of one against a unanimous majority. *Psychological Monographs*, 70(9): 1–70.

13 Deutsch, M., & Gerard, H. B. (1955). A study of normative and informational social influences upon individual judgment. *Journal of Abnormal and Social Psychology*, 51(3): 629–36.

14 Ariely, D. (2008). *Predictably Irrational: The Hidden Forces That Shape Our Decisions.* New York: HarperCollins Publishers.

15 Mazar, N., Amir, O., & Ariely, D. (2008). The dishonesty of honest people: A theory of self-concept maintenance. *Journal of Marketing Research*, 45(6): 633–44.

16 Wegner, D. M., Erber, R., & Raymond, P. (1991). Transactive memory in close relationships. *Journal of Personality and Social Psychology*, 61(6): 923–9.

17 https://blog.kaspersky.com/digital-amnesia-survival/9194/

18 Epley, N., & Whitchurch, E. (2008). Mirror, mirror on the wall: enhancement in self-recognition. *Personality and Social Psychology Bulletin*, 34(9): 1159–70.

19 White, D., Burton, A. L., & Kemp, R. I. (2015). Not looking yourself: The cost of self-selecting photographs for identity verification. *British Journal of Psychology.*

20 http://www.psy.unsw.edu.au/news-events/media/2015/07/study-we-dont-look-we-think-we-look

21 Harris, C. B., Keil, P. G., Sutton, J., Barnier, A. J., & McIlwain, D. J. (2011). We remember, we forget: collaborative remembering in older couples. *Discourse Processes*, 48(4): 267–303.

22 Vredeveldt, A., Hildebrandt, A., & Van Koppen, P. J. (2015). Acknowledge, repeat, rephrase, elaborate: Witnesses can help each other remember more. *Memory*, 1–14.

23 Skagerberg, E. M., & Wright, D. B. (2008). The prevalence of co-witnesses and co-witness discussions in real eyewitnesses. *Psychology, Crime & Law*, 14(6): 513–21.

24 Roediger, H. L., & Butler, A. C. (2011). The critical role of retrieval practice in long-term retention. *Trends in Cognitive Sciences*, 15(1): 20–7.

Chapter 9

1 http://www.boston.com/news/local/massachusetts/articles/1995/03/19/questions_prompt_reexamination_of_fells_acres_sexual_abuse_case?pg=full

2 Referenced in De Young, M. (2004). *The Day Care Ritual Abuse Moral Panic.* McFarland.

3 Summit, R. C. (1983). The child sexual abuse accommodation syndrome. *Child Abuse & Neglect*, 7(2): 177–93.

4 London, K., Bruck, M., Wright, D. B., & Ceci, S. J. (2008). Review of the contemporary literature on how children report sexual abuse to others: findings, methodological issues, and implications for forensic interviewers. *Memory*, 16(1): 29–47.

5 http://www.wicca-chat.com/bos/witch/amiraults-trial.txt

6 https://www.nspcc.org.uk/preventing-abuse/signs-symptoms-effects/

7 Kendall-Tackett, K. A., Williams, L. M., & Finkelhor, D. (1993). Impact of sexual abuse on children: a review and synthesis of recent empirical studies. *Psychological Bulletin*, 113(1): 164–80.

8 Pazder, L., & Smith, M. (1980). *Michelle Remembers.* New York: Pocket Books.

9 Loftus, E. F., & Guyer, M. (2002). Who abused Jane Doe? The hazards of the single case history. Part I. *Skeptical Inquirer*, 26(3): 24–32.

10 Webster, R. (1995). *Why Freud Was Wrong: Sin, Science, and Psychoanalysis.* Basic Books.

11 http://www.nobelprize.org/nobel_prizes/facts/literature/

12 http://theconversation.com/explainer-what-are-false-memories-49454

13 Patihis, L., Ho, L. Y., Tingen, I. W., Lilienfeld, S. O., & Loftus, E. F. (2014). Are the 'memory wars' over? A scientist-practitioner gap in beliefs about repressed memory. *Psychological Science*, 25(2): 519–30.

14 Cheit, R. E. (2014). *The Witch-Hunt Narrative: Politics, Psychology, and the Sexual Abuse of Children.* Oxford University Press.

15 Hebl, M. R., Brewer, C. L., & Benjamin Jr, L. T. (eds.). (2001). *Handbook for Teaching Introductory Psychology*, Vol. 2. Psychology Press.

16 http://www.innocenceproject.org/

Chapter 10

1 Hart, J. T. (1965). Memory and the feeling-of-knowing experience. *Journal of Educational Psychology*, 56(4): 208–16.

2 Eakin, D. K., Hertzog, C., & Harris, W. (2014). Age invariance in semantic and episodic metamemory: both younger and older adults provide accurate feeling-of-knowing for names of faces. *Aging, Neuropsychology, and Cognition*, 21(1): 27–51.

3 Jaeggi, S. M., Buschkuehl, M., Jonides, J., & Perrig, W. J. (2008). Improving fluid intelligence with training on working memory. *Proceedings of the National Academy of Sciences*, 105(19): 6829–33.

4 Melby-Lervåg, M., & Hulme, C. (2013). Is working memory training effective? A meta-analytic review. *Developmental Psychology*, 49(2): 270–291.

5 Melby-Lervåg, M., & Hulme, C. (2015). There is no convincing evidence that working memory training is effective: A reply to Au et al.(2014) and Karbach and Verhaeghen (2014). *Psychonomic Bulletin & Review*, 1–7.

6 Foer, J. (2011). *Moonwalking with Einstein: The Art and Science of Remembering Everything*. Penguin Books.

7 Geraci, L., McDaniel, M. A., Miller, T. M., & Hughes, M. L. (2013). The bizarreness effect: evidence for the critical influence of retrieval processes. *Memory & Cognition*, 41(8): 1228–37.

8 Morrison, K. M., Browne, B. L., & Breneiser, J. E. (2012). The effect of imagery instruction on memory. *North American Journal of Psychology*, 14(2): 355–64.

9 https://www.ted.com/talks/elizabeth_loftus_the_fiction_of_memory/transcript?language=en

Index

C